THE WEST TEXAS POWER PLANT THAT SAVED THE WORLD

REVISED & EXPANDED EDITION

THE WEST TEXAS POWER PLANT THAT SAVED THE WORLD

ENERGY, CAPITALISM, & CLIMATE CHANGE

ANDY BOWMAN

TEXAS TECH UNIVERSITY PRESS

This book is typeset in EB Garamond. The paper used in this book meets the minimum requirements of ANSI/NISO Z39.48-1992 (R1997). ⊗

Designed by Hannah Gaskamp
Cover illustration by Hannah Gaskamp

Library of Congress Control Number: 2023941095
ISBN 978-1-68283-186-1 (paperback)

Texas Tech University Press
Box 41037
Lubbock, Texas 79409-1037 USA
800.832.4042
ttup@ttu.edu
www.ttupress.org

For my mother Phyllis and my grandfather Paul

CONTENTS

Ours is the century in which all man's ancient dreams—and not a few of his nightmares— appear to be coming true.

ARTHUR C. CLARKE, *REPORT ON PLANET THREE AND OTHER SPECULATIONS*, 1972

FOREWORD

"Climate change is normal, natural, and necessary," a Houston man lectured me on Twitter. Another shared a homemade graph purporting to contain ice core data from Greenland that buttressed his argument that the planet isn't warming.

As a climate scientist, I find my inbox and social media feeds are regularly inundated with objections to the science I study. Do these objections have any basis in reality? The answer is, sadly, no.

Burning coal, gas, and oil provides us with energy. We use it to electrify our homes and our cities, power our cars and our planes, and fuel our industry and our manufacturing. For the last two hundred years, the vast majority of our energy has come from fossil fuels. However, their extraction and combustion produce massive amounts of carbon dioxide and other heat-trapping gases. As these gases build up in the atmosphere, they are essentially wrapping an extra blanket around the planet, causing it to warm.

These facts have been well understood for a very long time. Scientists connected fossil fuels to carbon dioxide, and carbon dioxide to global warming, in the 1850s. By 1965, scientists were sufficiently concerned about the risks of this rapid, human-caused warming to formally warn a US president. And by 2020, after decades of research and thousands of scientific studies, scientists confirmed they were 99.9999 percent confident that humans were responsible for all the observed warming. It doesn't get any more certain than that.

Given how old and how solid the science is, when we hear people question it, we often assume they don't know the facts. And if that's the case, then the answer is clear: we need more and better education to get them on board. Anyone who understands the science, we think, will take this problem seriously and act quickly.

Does this approach work? In low-income countries with few fossil fuel resources and minimal heat-trapping gas emissions, more education does make people more concerned. And it should! But in rich countries like the US, blessed with plentiful coal, oil, and gas resources, we have contributed far beyond our fair share of heat-trapping gas emissions.[1] We know this instinctively, even when we try to deny this knowledge or ignore it. So it's no surprise that more information on how serious climate change is can be polarizing. For those already worried, it can make us even more anxious—even panicked. But if we had already decided we weren't concerned, then our defenses tend to engage even more strongly. Rather than changing our minds, often this just motivates us to seek out science-y sounding objections—like the examples from social media I shared above—to justify our position more ardently.

The reality is that such objections are simply a smoke screen that obscures the real reasons we balk at accepting the reality of a changing climate and our own contributions to this warming. These reasons are known as *psychological distance* and *solution aversion*. In plain English, we don't think climate change matters to us, and we don't think there's anything we can do to fix it.

Psychological distance is a phenomenon to which we humans are curiously prone. We have an inborn tendency to perceive many risks as being far away from us: in time, in space, and in relevance. We pile up debt on our credit cards, ignore recommended nutrition guidelines in favor of unhealthy foods, and tell ourselves we'll exercise tomorrow. Why? It's not out of any doubt that debt is real, heart disease can kill, and exercise is beneficial. It's because we don't see the risks as imminent. Climate change demonstrably invokes these avoidance mechanisms. Across the US, 72 percent of people agree that climate is changing, but only 43 percent believe it will harm them personally. Because we view this as a distant issue, it is all too easy to push climate change to the back burner. The more scientific warnings we ignore, the more numb we become.

Andy Bowman lives in Texas. That's where I live, too. We know our state is naturally prone to more extreme, damaging climate and weather disasters than any other state. We've lived through many of these events ourselves, and they've left their mark on our memories: from the hurricane with which Andy begins his story to the massive dust storm that rolls over our town as I write.

Since 1980, Texas has been hit by 129 disasters whose damages topped $1 billion each. And that's exactly why Texas is also the most vulnerable state to climate impacts. Global warming is loading Texas's already weighted weather dice against us. It is making our heavy downpours more frequent, our droughts longer and more intense, our heat waves more deadly, and our hurricanes much stronger and more damaging. It's estimated, for example, that human-caused warming of the planet was directly responsible for nearly 40 percent of Hurricane Harvey's record-breaking rainfall and *three quarters* of the more than $100 billion in economic damages the storm caused.

Even more concerning, however, is solution aversion—the belief that the cure is worse than the disease, so to speak. In February 2021, we experienced this in real time. A massive winter storm—a vortex of cold air from the Arctic—covered the entire state in ice, snow, and record-breaking cold temperatures. It left millions of people without electricity, heat, or water, for days and even weeks, following the disaster. Our state's response was a classic example of both psychological distance and solution aversion. To start with, these freezes have happened before: just ten years ago, as a matter of fact. After that freeze, reports and recommendations were made, yet none were implemented. Why not? The risks were viewed as distant. Who knew when the roll of the dice would bring back such a rare, unusual event, legislators reasoned. Then, as power outages spread across the state, solution aversion kicked in as well. Rather than acknowledge the fact that the earlier recommendations hadn't been heeded, Governor Greg Abbott went on television to blame frozen wind turbines. But when all the numbers were crunched, it turned out that a full 87 percent of the power outages were due to failures to winterize natural gas plants, with a small contribution from outages at nuclear plants. Wind accounted for only 13 percent of the outages. And even there, it was because Texas producers had declined to install the winterization technology that keeps them turning in much colder temperatures. Turbines from Antarctica to the Arctic have this technology and don't freeze up.

From Hurricane Harvey to the winter storm of 2021, extreme events happen with or without climate change. But climate change is making them worse: stronger, bigger, longer, and sometimes more frequent, too. Even polar vortexes are affected, as Arctic warming makes the jet stream—which holds this cold air up north—slow down and begin to meander, allowing cold air extrusions to reach further south. As Texans, we already get enough extreme weather; we can't afford any more.

When we start to talk about climate change solutions, though, we don't only confront legislators' reluctance to prepare for rare but extreme events. We also run head-on into the fact that Texas is the number one producer of carbon pollution in the United States. If Texas were its own country, it would be the seventh largest emitter in the world. That puts it ahead of Iran, South Korea, and even Canada. There's no getting around it: Texas is a big part of the problem.

But Texas is also the state with the greatest potential for solutions. Texas is already, and has been for a long time, the leading wind producer in the United States. Wind eclipsed coal in Texas in 2020, producing nearly 23 percent of the electricity on the ERCOT grid. Then there's solar: when I moved to Texas almost fifteen years ago, the state was not even on the list of top ten solar producers. As of 2020, Texas is number two in the country, and the state's solar capacity is slated to double in 2021. Soon, Texas will top that list too, and in this book, Andy Bowman explains exactly how this has happened and why it's such good news.

In *The West Texas Power Plant That Saved the World*, Andy dismantles the concepts of psychological distance and solution aversion. With vivid stories and incontrovertible facts, he makes it clear that climate impacts are not a future problem; they are here and now, affecting all of us in ways that matter. But he also explains how solutions are here today as well. They do not involve a return to the Stone Age or a complete destruction of our energy-intensive way of life. Instead, from home-grown Texas solar to far-off Chinese investment, the world is already changing. Clean energy is already here. And the future can be bright.

That's why this book is so powerful. From his uniquely and intimately personal perspective, Andy shows how climate change is not a distant issue. He hails from Galveston and now lives in Austin, so he has experienced hurricanes and heat waves amplified by climate change. He understands the stakes, and his experience working in the renewables sector over the last twenty years has shown him that Texas can also be a big part of the solution, because Texans understand energy. The state has its own electricity grid, and in this book, Andy eloquently explains how that has both helped and hindered the state's embrace of green energy. Here in Texas, we have a lot of sun and a lot of wind. That uniquely positions us to become leaders in clean energy, a space that, as Andy clearly explains, is already being ruled by China today. As Andy lays out, we must catch up to China, which is already doing more than the United States, from both a clean-energy technology and a carbon market approach.

The bottom line is clear. Climate solutions are not harmful, they are not job killers, and they are not a drag on the economy. Climate solutions are energy solutions. Climate solutions grow jobs. Clean energy already powers everything from the Dallas–Fort Worth airport to the largest Army base in the United States, Fort Hood in Killeen, Texas. Clean energy makes us more competitive at the international scale. And clean energy springboards us into a better future where Texas can continue to lead the world in energy, but not at the expense of the quality of its air, water, and its people's health.

Climate change isn't only an economic challenge: it represents a great economic opportunity to wean ourselves off our old and dirty ways of getting energy and replace those with the clean renewable energy sources we can grow right here at home. Yes, Texas is part of the problem—but as this book explains, Texas is also a big part of the solution. And if Texas can fully embrace renewable energy, can't we all?

KATHARINE HAYHOE

Chief Scientist, The Nature Conservancy
Paul Whitfield Horn Distinguished Professor and
Political Science Endowed Chair, Texas Tech University

THE
WEST TEXAS
POWER PLANT
THAT SAVED
THE WORLD

PART 1

THE POWER PLANT THAT SAVED THE WORLD

CHAPTER 1

THE GHOST OF INDIANOLA

Galveston's infamous 1900 Storm, and the city's response to it, have much to tell us about our inclination to think we're managing climate change better than we really are.

J ust after midnight on August 17, 1983, an unusual hurricane "drought"—more than three years without a single storm making landfall on the continental US—broke with a vengeance on Galveston Island's west end. I was fifteen at the time and, having grown up steeped in Galveston's rich but traumatic history of hurricanes, was not going to miss it. Late that night, my best friend and I quietly took our bikes from the garage and rode out into the flooded streets and wailing wind. Some kids sneak out to parties at night; we snuck out to see the hurricane.

Hurricane Alicia had been born days earlier when a faint eddy in the upper atmosphere spun off an odd August cold front. This slight convection sat over the warm Gulf of Mexico long enough to brew a tropical depression, which quickly grew into a tropical storm and then a hurricane. Steering winds pushed it toward the coast and just prior to landfall, as if making up for the last few years' missed storms, Alicia blossomed unexpectedly into a Category 3 major hurricane.

Normally I would have been prevented from doing something so foolish as going out that night, but my father was asleep, and my mother had failed to convince me the day before to evacuate to her house in Houston. I also thought, wrongly, that Alicia wouldn't be much different from all the other tropical weather I witnessed growing up in Galveston. Fierce storms with heavy rain, thunder, and lightning that flooded our neighborhood happened all the time on the island. In fact, my friends and I would often bodysurf in the big waves from a tropical disturbance and venture out after storms to paddle the streets in canoes or surf behind a bicycle.

We could just manage to ride our bikes through the foot or so of water covering the streets, stopping every couple of blocks to look around in wonder. Powerful sheets of rain bore down across the dark neighborhood, brightened occasionally like daylight by lightning spidering across the sky. In the midst of Alicia, Galveston was an alien world. Not a person in sight, rain almost too dense to see through, wind shrieking and howling, and the water in the streets eerily ocean-like, whipping into small whitecaps where cars normally drove.

We finally approached the main event: Galveston's Seawall Boulevard, the street running atop the giant wall built nearly a century before to hold off storms just like this one. Here was ground zero of the war between the furious ocean and the island. We left our bikes and trudged up to the side of a building shielding us from the wind, took a last breath, and then ventured into the full-bore hurricane. The raindrops, warm like the summer ocean, felt like blows against our bodies, and we could only barely hold our ground against the wind. It was frightening, but even more exhilarating, to experience the storm at its height. Just then, a piece of debris—a board or part of someone's roof—whizzed by my arm, and I suddenly realized what I somehow had not before: that the wind could carry more than just rain, something that could do us great harm. Having survived a few moments, we decided to call it quits. We took one last look at the enormous waves blasting up against the seawall, got back on our bikes, and made it home to safety.

I often reflect on that night and how easily one of us could have been hurt or killed, and what a senseless death it would have been; as a parent today, I am utterly appalled at myself. That said, I will never forget the experience of peering beyond the seawall and witnessing the full fury of the hurricane firsthand.

By the next afternoon, the storm had passed and the sky was blue, but the damage was unbelievable. Houses and cars everywhere were flooded and debris several feet high lined the streets. The power and water were out and would be for days. Along the seawall, a large video-game arcade built over the sand—a building in which we had spent countless hours in prior summers—was completely gone; only the poles on which it had rested remained. The National Guard was then deployed, and soldiers patrolled the streets with machine guns draped over their shoulders. This seemed like overkill to me, but looters had been on the streets since the morning after. A sunset curfew was imposed and with the power out, there was nothing to do but explore the damage by day and listen to the mosquitoes buzz in the hot air at night. The next day, leaving the island for my mother's place in Houston, we passed boats, parts of houses, chairs and debris of all kinds—perhaps even including pieces of our favorite arcade—for what seemed like miles resting atop the freeway.

■

My obsession with Mother Nature's wrath—precursor to my eventual career in renewable energy and my personal frame of reference on climate change—all began with Galveston and its hurricanes.

The small city of about 50,000 people sits on the eastern tip of the twenty-seven-mile-long barrier island, its back to the mainland and its gaze far out to sea. Behind it lies the rich and fertile coastal prairie inclining to the Great Plains and Chihuahua Desert; before it lies more than 650 miles of open ocean, the same distance to Kansas City, Missouri, in the other direction.

Geography being destiny, Galveston's perch at the edge of the warm and restless Gulf of Mexico has shaped its history, both for better and for worse. First settled by Karankawa and Akokisa tribes who fished and oystered its rich shores, it was established as a trade center by French, Spanish, and Mexican settlers and then a pirate base of operations before becoming one of the busiest cotton ports in the American South. In the 1920s, its bustling port and ready access to international waters made it an ideal bootlegging, gambling, and debauchery destination, earning it the nickname the "Free State of Galveston." During World War II, US Navy ships launched from Galveston prowled the Gulf for menacing German submarines. These days, the island is mainly a tourist destination where beachgoers suntan as nearby tankers line up to move through the Houston Ship Channel, one of the busiest shipping lanes in the world.

Punctuating its remarkable history like a drumbeat is the march of terrible hurricanes rising up from Gulf waters to strike the island. Galveston, it seems, is never far from the next storm: on average, hurricanes with winds over a hundred miles per hour strike the island once every nine years. Island cities sit on civilization's front porch, wide open and exposed to the elements like nowhere else.

Of all the storms that have struck the island, one towers above all the rest, the simply named "1900 Storm" that devastated the city so completely that its very survival has come to define it more than anything else. Likewise, the ordeal of the storm, the community's epic response to it, and the century of storms that have come behind it have marked us Galvestonians with a special view on the question of mankind's ability to manage natural phenomena.

On September 8, 1900, an unexpected gale arrived late in the day and that night, in pitch-black darkness, the storm's 145-mph winds and fifteen-foot storm surge overwhelmed the island, which stood not even eight feet above sea level at its highest point.[1] At some time during the night, the steel streetcar tracks that ran beachside across the length of the city were uprooted, turned sideways, and then pounded forward by the raging waves, picking up more debris as they went. This makeshift battering ram scraped the ground clean where city blocks had stood the day before. Among the countless heartbreaking stories of that devastating night, none is more tragic than the ten nuns and ninety-three orphans of St. Mary's Orphan Asylum. The nuns and children took shelter in the two-story building, but as the waters continued to rise, the building collapsed and the group was forced out into the raging black waters. One of the nuns tied herself, using bedsheets, to the ten orphans in her charge; they were all found together buried in the sand the next day.[2]

As quickly as the storm had appeared, it then passed further inland. By late the next morning, the rains stopped and the angry waters of the Gulf had eerily calmed to nearly normal.[3] A city of 38,000 at its peak both economically and politically on the eve of the storm, Galveston was cut down to rubble literally overnight. Photographs of the aftermath more closely resemble Hiroshima or Dresden than any storm-damaged town. In all, the storm claimed the lives of between six thousand and twelve thousand people; so massive was the destruction that the exact number of dead has never been determined. To this day, the 1900 Storm is the deadliest natural disaster in US history.[4]

In the aftermath, as Galvestonians began to pick up the pieces, there was much discussion about what should be done in light of the certainty of future storms. Everyone in Galveston at that time knew the story of a different town, just down the coast, and a different storm. Fourteen years earlier, a fierce hurricane had struck the small port town of Indianola, which had only recently recovered from a prior direct hit in 1875. Indianola was a growing commercial center just beginning to compete with Galveston for shipping, but the 1886 hurricane changed everything. The Indianola Storm was a monster, the fifth strongest ever to strike the United States, and its 150-mph winds leveled almost the entire town. Adding insult to injury, hours after the storm passed a fire started that burned most of what remained. Unbelievably, five weeks later yet another hurricane came ashore down the coast, close enough to flood Indianola once again. Understandably, Indianolans had had enough: the town was simply abandoned, and the surviving residents dispersed to other places. Today, Indianola's remains can be found resting under Matagorda Bay, within view of a lonely historical plaque.

Storms had come and gone as long as anyone on the island remembered, but just as Galveston plunged forward into the new and modern twentieth century, it had been reminded of something terrifying and timeless. What Indianola represented was the idea that a big enough storm coming at the wrong time and place could deal a truly mortal blow to a coastal town. This would not be Galveston's fate, city leaders decided. As Galveston worked to recover from the storm, it built its resolve not to let the storm, nor any future storms, vanquish it. As a result, the effort to rebuild after the 1900 Storm came to rival the intensity of the storm itself. The city not only completely rebuilt its damaged and destroyed buildings, roads, and bridges but also embarked upon not one but two massive infrastructure projects, each of such a scale that it must have seemed at the time like something out of a science fiction novel.

First, to protect against future storm surges, a seawall standing seventeen feet high was planned along three miles of the Gulf side of the island. A special board of expert engineers from around the country devised the extraordinary plan, which accounts at the time described as "one of the most stupendous schemes of protection and rehabilitation that has ever been attempted on the engineering stage."[5] Construction started in 1902 and lasted two years. First, massive anchoring timbers were pounded through the sand into the clay far underneath; then the seawall was poured, weighing about 40,000 pounds per foot; behind it, a one-hundred-foot-wide embankment was constructed.[6] Expanded several times over the decades as the city grew, Galveston's seawall today is almost ten miles long and is said to have created the longest sidewalk in the world.

The second project was even more ambitious: Galveston raised the grade, by several feet, of the entire populated portion of the island. The new seawall could protect against violent waves, but alone it would only get half the job done; only by literally raising the island behind it up to the same height could it be made secure from future floodwaters coming in from the bay side of the island. For this massive undertaking, about twelve million cubic feet of sand dredged from the bottom of Galveston Bay was pumped in to raise four square miles of land. An account at the time noted that the amount of sand was enough "to build five Egyptian pyramids as large as the famous Cheops."[7] This portion of the island was raised by an average of four feet, although most residential areas were raised about eight feet and some areas were lifted by as much as seventeen feet.[8] As the city later grew, and as late as 1950, additional parts of the island were similarly raised.[9]

Even by today's standards, Galveston's grade raising is on a scale difficult to imagine. The most populated part of the island was cut open with a new man-made canal that provided access to the dredge boats ferrying in spoils from the bottom of Galveston Bay, raising the ground block by block, neighborhood by neighborhood. Completed over eight years by an army of engineers and laborers, the project raised five hundred city blocks, including the water, sewer, gas, and power infrastructure systems in their entirety, and more than two thousand structures.[10] By the time the canal itself was refilled, the total cost at the time was $2 million, or about $54 million in 2016 dollars.[11] Executing such a project today, using current construction norms, would dwarf these figures by a vast order of magnitude.

Standing on top of the seawall that night in 1983 during Hurricane Alicia, I could easily imagine how terrifying the 1900 Storm must have been without it. The following day, looking down on the waves gently lapping seventeen feet below and then turning around to see the damaged, but still standing, homes, churches, and restaurants that had been raised out of harm's way, I felt a tremendous sense of safety and security. It could have been so much worse, had the seawall not been there to prevent the waves from washing over the entire island again. I felt something akin to the awe Galvestonians must have felt in 1904 upon the seawall's completion. What an incredible job they had done. It seemed truly inconceivable that circumstances could ever conjure enough waves and water to reach all the way to where I stood.

■

When my family first moved to the island in the early 1970s, Galveston's economy was ailing and the beautiful homes in the historic East End were as neglected as they were cheap. My parents bought a two-story Victorian home with high ceilings, giant windows, an inoperative fireplace, and transoms above each door, and set about restoring it. Among the house's charms were its wavy floors. As a child, I would sometimes place a marble on the floor in one corner of my bedroom and watch it roll left and right, meandering all way to the far corner. I wondered why this was so, until one day an old man accompanied by a cameraman knocked on our door. He was famous filmmaker King Vidor, and he asked permission to film the house for an autobiographical work. It turned out that as a six-year-old boy, Vidor had survived the 1900 Storm in our house, and in my very room. His stories of the night of the storm, when waters reached the second-floor windows, and its aftermath, were vivid and terrifying.[12]

For those growing up in Galveston, the 1900 Storm was omnipresent. We learned about it at school, heard stories about it from survivors, and walked streets with historical markers detailing it here and there around town. One could not avoid the photos, books, exhibits, and films about it everywhere. Every schoolchild knew that the day before it arrived, Galveston was a city of prospects, wealth, power, and apparently limitless growth, comparable to New Orleans in terms of economic, intellectual and political activity; and we all knew the terrible story that comes after.

Summers on the island meant keeping tabs on hurricanes that might be bowling through the Atlantic at us while spending most of our days swimming at the beach or fishing, crabbing, or exploring. Island winds were almost always clear ocean air, but occasionally winter cold fronts blasted in from the north, bringing with them air from the highly industrialized mainland. Galveston has practically no heavy industry, but to its back is the largest petrochemical complex in the world, stretching from the plants outside Baton Rouge to Texas City's sprawling smokestacks directly across Galveston Bay and on down past Dow Chemical's vast facilities at Freeport, where a good job first brought my grandfather decades before. Those rank north winds made it easy for Galvestonians to appreciate the importance of clean air and water, and I grew up an authentic if naive environmentalist.

I went away to high school and college in the Northeast and then to Austin for law school, where I learned that "environmental law" meant mainly permitting power plants. So, a couple of years out of school and hoping to do some good for the world, in 1998 I took a job as general counsel of a start-up wind-power company. Our prospects were questionable at best: at that time all of Texas had only a handful of wind turbines operating, and I knew precious little about how to practice law, much less be a general counsel. That same year, my wife and I bought a home we could barely afford and a month later had our first child. Laden with debt, significantly underqualified for my job, and working for a start-up in an industry that barely existed, I view these momentous decisions in retrospect as audacious votes for the future.

Our timing at the wind start-up was good, however, and I was able to quickly learn core tasks like drawing up land leases, grid interconnection agreements, and power purchase contracts. We were fortunate that the price of oil had collapsed in the late 1990s, so no one in windy West Texas had much to lose by talking to us about building wind projects on their land. Most of the county officials, landowners, and utilities we dealt with on these early wind projects couldn't decide if we were seriously planning giant wind turbines to sit on top of West Texas mesas or if maybe we were up to no good somehow, given our outlandish-sounding plans. Windmills weren't anything new in West Texas, but wind turbines 150 feet high with blades eighty feet long, and costing a million dollars each, definitely were. Few seemed to believe we would follow through with the projects we proposed, but we did, and over a few years' time we succeeded in finding partners to get several projects built. One of these projects, the King Mountain wind project in Upton County, was, at 280 megawatts, the largest wind project in the world when completed and held that record for many more years.

A couple of years later, I left that position and started my own wind company with two friends. We crisscrossed the country in the early 2000s, from Oregon and California to Michigan and New York, talking with landowners and state and local officials and utilities about the new wind energy technology and the projects we wanted to build. We struggled for years, but the company was successful and in 2005 it was purchased by an Irish wind energy company, Airtricity. In 2008, Airtricity was purchased by an even bigger company, E.ON Climate & Renewables, the clean-energy affiliate of a large German utility. Each year with these companies, we built more and more wind projects, from Texas to New York and Pennsylvania. I managed a growing project development scope across these companies as they grew, eventually as the chief development officer for E.ON North America. A couple of years later I left E.ON and started another company, Pioneer Green Energy, where my partners and I developed new wind and solar projects across the country from California to Maryland. Amid all this, my wife and I had our second and third children, paid off our debt, and moved into a new house.

Somehow, over a couple of decades, our children have nearly grown up and this work has become my career. I have spent almost twenty-five years working in renewable energy and recently started another new company, this time in energy storage. All the wind and solar projects I helped develop were large, utility-scale power plant projects, meaning they had to live or die fighting for a place in wholesale electricity markets dominated by traditional power plants—mainly coal and natural gas. Many of the projects we tried to develop were not ever built, but many were successful and are operating today. All told, over my career I developed myself and with my team, or was responsible for developing, wind and solar projects totaling about 3,600 megawatts. To give a sense of how much generation this represents in traditional power plants, it is equivalent to the generating capacity of about seven average-size coal power plants. In the context of the renewable energy business, these projects represent about 2.5 percent of all the wind and solar built in the US through 2017 (144 gigawatts).[13] I am proud to be associated with this quantity of clean energy, which I felt did measurable good, displacing polluting generation that otherwise would have been built.

The irony that the same sun and wind I had treasured in my youth on Galveston Island had become my full-time job as an adult never escaped me. If geography is indeed destiny, as it so clearly has been for Galveston, the same is surely true for people. The island's long history of storms made it what it was, and me much of what I am—and although I no longer live there, I remain connected to it still and always miss its warm and salty breeze. This point became clear to me in 2008 when events back at Galveston conspired to remind me of the other face of Mother Nature, which put into a very different perspective the work I had been doing in clean energy.

■

On September 13, 2008, five days after the 108th anniversary of the 1900 Storm, Hurricane Ike struck the island. Although only a Category 2 storm at landfall, Ike wreaked breathtaking devastation on Galveston, damaging 80 percent of all structures on the island and flooding almost every street and neighborhood. Ike tracked the 1900 Storm's path almost exactly, its most powerful northeast quadrant raking directly across the most populous part of the island. Just as the 1900 Storm had done, Ike's astounding twenty feet of storm surge inundated the island from both sides, this time with waves leaping over the seawall and floodwaters simultaneously rising up from the bay behind. Ike left six feet of water inside City Hall and, incredibly, ten feet or more in homes across many low-lying neighborhoods. The beach town of Bolivar, sitting on a thin peninsula across the bay from Galveston Island, was all but bulldozed into the water by the pounding waves and wind, in much the same way Galveston had largely been in 1900.

Vast parts of the mainland were similarly devastated by Ike. Houston's metropolitan-area population had grown all the way to the coast over prior decades, with many new subdivisions along the water and in lower-elevation areas near interior waterways, all severely damaged. Ike blew glass windows out of Houston skyscrapers and flooded numerous parts of the city and metropolitan area worse than anyone could remember.

By this time I was living in Austin, but I returned to Galveston to help clean up my relatives' badly damaged homes. Receding floodwaters left a car resting against the front door of my aunt's home; we entered through the rear and inside, I reached up to touch the high-water mark, a straight line almost to the ceiling across her living room wall. It was the same in every other home you could see for blocks in all directions. My grandmother lived across town, just behind the seawall in one of the higher parts of the island, and she had about two feet of standing water throughout the house.

On a break from shoveling debris onto the street at my grandmother's, I took time to walk up and inspect the mighty seawall. It was the same towering concrete, undamaged, but this time I surveyed wreckage on all sides much worse than that left by Alicia or any other storm I had seen. The biggest waves had smashed over the wall, sending ocean water careening down the other side. Piered restaurants that had extended out over the beach days before had vanished; random orphaned boats had smashed over the seawall and into buildings along Seawall Boulevard; cars had been swept this way and that mixed in with the boats; debris of all kinds lay strewn across the street and throughout city blocks behind it.

Considering the damage Ike had caused, not only in Galveston but across the whole Upper Gulf Coast, it suddenly became clear to me that Galveston's twin super projects completed in the wake of the 1900 Storm, that I had always been so enamored with, had not in fact outsmarted the hurricanes. Perhaps a town the size of Galveston in 1900 might have successfully hidden from a storm like Ike from behind the seawall; perhaps it would have suffered only significant, not disastrous, harm. But the seawall had barely grown over all those years, while the population of Galveston had almost doubled. More importantly, the greater Houston/Galveston area population had grown massively, most of it with no seawall to protect it. In 1900, Houston had about 44,000 residents, barely larger than Galveston; by the time Ike struck, the Houston/Galveston population was nearly six million people[14]—essentially, one giant, sprawling population center extending more than an hour's drive from the coast and longer than that from side to side with no end in sight. By 2040, the total area population is expected to nearly double to ten million.[15]

As I stood once more atop the seawall, this time in the aftermath of Ike, suddenly its symbolism to me inverted. In that moment, the seawall and the grade raising—perhaps the most ambitious geoengineering efforts ever undertaken in this country to protect against storms just like Ike—transformed from Galveston's protective talisman against storms into a symbol of the smallness of our efforts to protect ourselves from the ballooning risks we face today from climate change. Galveston's heroic engineering feats had, years before, impressed me for how many they protected; now they seemed most remarkable for how very many more they do not.

This same lesson, more or less, had played out a few years earlier with Katrina in 2005, except that there it wasn't a seawall that failed a community but its extensive dyke system. It happened again in 2012 with Sandy, and then again, of course, in 2017 with Harvey, Irma, and Maria. Harvey—the greatest single rain event ever recorded in the lower forty-eight states—also remains a contender for the most expensive storm in history in damage caused. As of this writing in late 2020, millions of Americans watch the weather for news of the next storm inevitably heading our way during hurricane season.

In that same moment, contemplating all Galveston's heroic efforts alongside Ike's devastation, I felt the ghost of Indianola pass through me. Perhaps it isn't Galveston, but instead the great-grandchildren of Indianola's refugees, living now in San Antonio, Tulsa, or who knows where, who have the last word about our efforts to prevail against nature's occasional wrath. Maybe it wasn't Indianola's failure to believe that the storms battering Texas's coast could be neutralized that we should have taken to heart from those pair of turn-of-the-century storms but rather Galveston's folly to think that even our boldest ingenuity ever could stop them.

We know now that climate change will bring even stronger storms than those that devastated Galveston, Houston, New Orleans, Puerto Rico, and many more places. We also know that the number of people in harm's way in coastal areas has grown exponentially and grows more with every year. The lesson appears to be crystal clear. In the end, it wasn't Mother Nature's wrath that Galveston's massive infrastructure projects had outsmarted: it was us.

These hard realizations fit with others I had been having then about my work in renewable energy and the growing problem of climate change. While a majority of Americans had been aware of the basic facts of climate change since the 1990s, for most it seemed to be a problem that would be solved in the same way that technology eventually solves all our other problems. Following the 2006 release of Al Gore's film *An Inconvenient Truth*, however, some started to understand the matter as it really is: an urgent and potentially intractable crisis, not just another environmental issue. After I saw that film, which explains in simple, compelling terms how we cause climate change and the ways it can be expected to intensify, I was chilled. Its description of the fundamental drivers and ceaselessly accumulating consequences took root in my mind, confirmed as they have been by an abundance of scientific inquiry since.

About 75 percent of the global greenhouse emissions driving climate change come from burning fossil fuels for energy,[16] and I was proud of my role in helping to execute so many clean-energy projects displacing these emissions to some degree. But anyone with a rudimentary understanding of the vast scale of greenhouse gases emitted each year knows that the world's collective efforts to reduce emissions have been—to put it mildly—massively inadequate. While wind and solar power have grown greatly over the past twenty years, they are nowhere near preventing enough fossil fuel emissions to make even a meaningful, much less a determinative, difference.

Worldwide, total CO_2 emissions have continued to rise each year without exception, when instead they must fall dramatically to limit warming. As detailed further in the next chapter, scientists recommend a maximum level of global atmospheric CO_2, as measured in parts per million (PPM), at 350, but levels have climbed steadily upward from 320 in the 1960s to 380 in the 2000s, right across 400 in 2016 and on up to 417 in 2020.[17] Instead of stopping or even slowing, global emissions accelerated over the prior two years, marking the largest annual increases in global CO_2 concentrations ever measured,[18] and in 2020 the

COVID-19 pandemic and economic downturn have just barely slowed the rise. Global average temperatures, meanwhile, have tracked ever higher almost every single year in lockstep with these rising emissions.

The bottom line is that, while the many wind and solar projects that have been built around the world in recent years have done a certain and quantifiable good, they do not, collectively, add up to much. This is not apparent looking backward in time: total installed renewable energy plants have grown dramatically, year on year, breaking new records like clockwork. However, the statement is absolutely true looking forward. Unquestionably, there is not nearly enough clean-energy infrastructure being built today to meaningfully alter the trajectory of greenhouse gas emissions—not by a long shot. In sportsman's parlance, the incremental approach to building renewable projects the past couple of decades has been like needing to hit a grand slam but bunting a single.

With this growing awareness, the self-satisfaction I experienced looking out over completed wind and solar projects that I had worked on began to feel hollow. Hearing the wind rush by beneath turbines in West Texas and seeing the sunlight glint off solar panels in our projects in California and Maryland felt oddly and uncomfortably familiar; somewhere, I had definitely had this feeling before.

Just as I had come to see Galveston's seawall as but a fleeting victory against the hurricanes that have battered the Texas Coast from time immemorial—increasingly inadequate as the storms get stronger and the coastal population has ballooned—now I was making a similar realization about all the wind turbines and solar panels I worked to erect here and there. They make us feel as though we are solving a problem that in fact we are not and that actually is growing larger. Just as Galveston lived with a false sense of security behind its seawall while the larger threat grew, so are we now mistaking the paltry steps we have taken to date to address power plant emissions and the other causes of climate change as even remotely adequate.

■

I was preoccupied with these gloomy thoughts when I learned of the Barilla Solar project, the small solar power plant built in 2014 which is the subject of this book. (To be clear, Barilla is a project with which I have had no professional involvement whatsoever.) The more I learned about this project's unique nature, and the more I considered it within the context of the problematic convergence of the power business and climate change that I was coming to fully grasp professionally, the clearer it became that this power plant represented something much more significant than appeared at first glance.

It is obvious that incremental progress in addressing greenhouse gas emissions will not fit the bill, that something transformational will be required. This book is about the idea that something transformational has already begun, something most of us have not even much noticed, but something that represents an agent of powerful and positive change.

It turns out that even without an effective approach in place today to manage power plant emissions—the largest source of greenhouse gases by a mile—clean energy has emerged as a winning competitive technology with a durable future. This outcome is the result of the cliff-diving price curve of photovoltaic (PV) solar over the past two decades, which itself arose from a curious and surprising sequence of events spanning three centuries and culminating in a particularly dramatic transformation since 2008. The public at large seems to have a limited understanding—perhaps even a complete misunderstanding—of the important changes already occurring with this fundamental shift all around the world. Even less understood is the tectonic transformation of the electricity business that these changes foretell, their unique potential to transcend political obstacles to climate progress, and the many ways they can enable us to better mitigate and adapt to the alarming climate changes that are already well under way.

While the challenge of responding effectively to climate change seems overwhelming, we have somehow, almost in spite of ourselves, arrived at a critical and favorable point of inflection in our energy practices. This key moment is actually one of a handful of fundamental inflection points in the long history of people and energy, but one that could perhaps be occurring in time to prevent the worst of the terrible crisis that our prior energy choice has put in motion. Ironically, as discussed in part 3 of this book, this path presents a *return* to sustainable energy practices for our species, rather than a departure into unfamiliar territory.

True understanding of complex things is often a struggle for perspective, and this is particularly so with a subject like climate change, with its long arc traversing centuries both behind and ahead of us and its broad potential to draw into adversity so many threads of human affairs. It is my hope that this book will help those searching for perspective on this challenging subject and for an understanding of how it fits into both our history and our future. A good start is to realize what Galveston did not: we must not mistake the substantial measures we have taken to date to address escalating emissions as anything more than what they are, which is woefully inadequate to the scale of the problems we face in the years to come.

CHAPTER 2

THE CLIMATE POINT
OF NO RETURN

Sensational headlines and dense studies obscure the simple math about
when devastating climate changes will become inevitable.

W hen it comes to climate change, many otherwise well-informed people I know have only a general understanding, paired with a foreboding, an undefined sense that the outcome will likely be worse than they think. The overall impression is that we will experience significant changes in weather—hotter temperatures, bigger storms, worse droughts—and, at some point in the future, this will generate distressing consequences, such as rising sea levels, loss of glaciers and sea ice, and extinction of certain species.

While this understanding is not wrong, it is incomplete, and its deficiencies are worth noting. In fact, the body of knowledge about much of what we can expect to happen, and when, is in fairly clear focus. While there remain significant unknowns, the broad outlines of what changes are coming when are well understood, along with a surprising amount of detail about several specific events we can expect.

Given the low level of awareness of these variables even among people who take an interest in the topic, and the proximity in time of some predicted events and their gravity, it may be surprising to know that this body of knowledge is not buried in a secure location somewhere, hidden away by government scientists. In fact, the information is startlingly accessible, thanks to the internet and decades of hard work by a small army of scientists. Through the efforts of the United Nations Intergovernmental Panel on Climate Change (UN IPCC) and the U.S. Global Change Research Program (USGCRP), a group of learned people around the world have coordinated their efforts and compiled their findings into convenient, short summaries, which are regularly updated. These resources are just a few clicks away (i.e., the IPCC's *Fifth Assessment Report*, 2015, at http://www.ipcc.ch/ and the USGCRP's *Fourth National Climate Assessment, Vol. I*, 2017, at https://science2017.globalchange.gov/).[1]

These helpful summaries notwithstanding, it can be difficult to wade through the detail about all the forecasted changes. The fact is, amid the sea of information available today about climate change, even the committed may struggle to get a handle on the most vital questions of all: When will the really harmful changes we should expect happen? When is the point of no return, the moment that it will it be too late to prevent at least the worst of the changes coming? Has this moment somehow already passed, or does it remain still in front of us? And if it does, how much time remains? What does the future hold for our children and grandchildren? As a father of three children, I personally have pondered this question regarding not just my own but those of other parents around the world, a large proportion of whom possess only limited means to manage the approaching challenges.

While there are a great many complexities and uncertainties about the timing of the changes that scientists predict will occur, it turns out that the question of when dangerous levels of climate change will be reached depends on who you are and where you live. For some, like the Native American tribes living on the Gulf Coast in Louisiana who are being overwhelmed by sea level rise, or Arctic people whose homes are built on permafrost that is increasingly thawing, the level of what they'd consider "dangerous" has already been reached. In any given place—like Houston, for example—those who lost their homes or the lives of loved ones to Hurricane Harvey would concur; but others who were protected by being able to afford a home outside the flood zone, or who had adequate insurance and resources to rebuild, may disagree. Both geography and socioeconomic status determine who is vulnerable, when, and where.

In order to identify a climate point of no return, it was necessary to choose a number. For better or for worse, the science generally converged on 1.5 or 2 degrees Celsius as the point of moving from changes impacting a minority to affecting a majority and becoming increasingly widespread. The UN IPCC assessment finds with "strong confidence" that the consequences of climate change will be worse as temperatures increase: specifically, that most risks rise from "moderate" to "high" as temperatures move from 1°C to 2°C, and then to "very high" beyond 2°C.[2] Temperature differentials over the past several hundred thousand years have generally not fallen outside of the range of 2°C or 3°C—meaning that the entire evolution of our species, including our spectacular run over the past ten thousand years or so, has occurred within these temperature bounds. Put another way, there is no precedent for human civilization thriving in a habitat defined by temperatures outside of this range.[3]

Based on this principle, signatories to the Paris Agreement undertook to hold "the increase in the global average temperature to well below 2°C above pre-industrial levels and . . . to limit the temperature increase to 1.5°C above pre-industrial levels, recognizing that this would significantly reduce the risks and impacts of climate change."[4]

Working backward then, if 2°C is the right target for a maximum level of warming, when might we expect to reach the level of cumulative emissions that will, generally two or three decades later, result in this temperature increase? There is a fair amount of scientific consensus about the answer to this question. The best research available indicates that about 790 gigatons of CO_2 and the CO_2 equivalent of other greenhouse gases is the maximum amount that can be present in the atmosphere before a 2°C temperature increase is at least 66 percent likely to occur.[5] Our past emissions total about 560 gigatons, meaning that we have about 230 gigatons left to emit before reaching this point.[6] This means of understanding the rate and timing of climate change is known as the "carbon budget," because it allows analysis based on how we will "spend" the emissions remaining before hard consequences become highly likely.

How long will it be before this remaining 230 gigatons is emitted? Future emissions may be higher or lower than current levels, but if we assume what seems uncomfortably possible at the moment I am writing this in 2020—that for the next couple of decades, emissions remain more or less unchanged from what they have been in the last couple of decades—we will emit all of these 230 gigatons by around 2033.[7] If we assume instead that measures are taken to moderate emissions, then we will get to 2037.[8] Considering that the global COVID-19 pandemic and the extended lockdowns it triggered around the world resulted in only a 6.4 percent reduction in 2020 global CO_2 emissions, and that China's net 2020 emissions fell by 10 percent early in the year before roaring back to a 1.5 percent net increase over 2019, the intense momentum behind emissions drivers is unmistakable.[9] If recovery after COVID is robust and we fail to uncouple economic growth from emissions, we will likely see an increase in cumulative emissions rather than diminishment, accelerating the timeline up from 2033.

Yes, you read that right: If we continue on more or less the same path we are on today, and even if we achieve moderate reductions in worldwide emissions, within the next two decades we can expect to reach the point at which we are assured that global average temperatures will in time increase to the point when scientists believe we will endure "grave damage" and carry "significant risks" forward. And there is no longer doubt regarding the 1.5°C increase—that amount of emissions is projected to *have already been reached*, in 2019.[10]

When you hear people acting as if their hair were on fire worrying about climate change, this analysis may well be what is on their minds. We are indeed careening toward this last barrier against the point-of-no-return no-go zone, and we are all but certain to go right through it. Beyond it, all the information we have suggests, extremely serious consequences will be all but locked in. Despite all the facts and figures that one reads about expected consequences, it may be difficult to conjure the kinds of personal changes we will all be experiencing as this critical juncture comes to pass, but they are quite easy to foresee if you consider the changes within the context of your own loved ones. For instance, my mother was born in the 1940s as the postwar world surged forward into the Pax Americana that fostered the unprecedented peace and prosperity of the past seventy-five years. In her later years, she may see a world on a dramatically different trajectory—a world, based on the state of affairs over the last few years at least, coming apart, heading into conflict around the globe and millions of people facing a horizon of increased hardship. Galveston, the island city where she raised her family, will be threatened not only by stronger and more damaging hurricanes but also by steadily rising seas, as will other coastal communities all over the globe.[11]

Our youngest child, fifteen years old as I am writing this, will have just turned thirty by the time the planet's ultimate crossing of the 2°C barrier is assured. The Austin, Texas, that she grew up in will already be hotter by then, going from just twelve days over 100°F each summer to more than thirty, and more susceptible to both drought and instances of extreme precipitation.[12] Her children, if she has them, will surely be considering moving north to cooler locations. They will undoubtedly be even more challenged by these cascading changes: their world will not be wondering whether jarring effects of climate change are coming but figuring out how to manage and endure changes already unfolding. A long list of things my daughter and so many of us take for granted today—such as food and water security and rare-to-nonexistent incidence of tropical diseases like malaria and dengue fever in Central Texas—will become increasingly fraught over her children's lifetimes. To this we can now add an increase in the likelihood of global pandemics like COVID-19. According to Harvard University's Chan School of Public Health, "Many of the root causes of climate change also increase the risk of pandemics."[13] The work of mitigating and adapting to climate change will be an inescapable, central fact of their generation's time.

For many hundreds of millions of others' daughters and sons around the planet, the situation will be much more daunting. The severity of challenges presented by the array of deteriorating conditions climate change will bring to their regions will be compounded by less competent and capable government, and the options open to them, their children, and their grandchildren to make necessary adaptations will be significantly constrained.

Sadly, the picture gets darker still. Up to the 2020 reductions due to the COVID pandemic, the most current information available is that we are not just failing to meet the emissions reductions set in the Paris Agreement but instead actually *increasing* emissions levels. Global CO_2 emissions rose by about 1.5 percent in 2017 and 2.1 percent in 2018, very large increases relative to prior years when they briefly plateaued and then grew by .6 percent in 2019.[14] The driver of these increases was greater fossil fuel usage in China and India, which more than offset slight reductions in the United States and Europe.[15]

In 2015, our family traveled to Delhi and Beijing, where we witnessed the reality of these sky-high and climbing emissions firsthand, seeing the effects of the air pollution that fossil fuel combustion creates alongside the heat-trapping gases that drive climate change. While an average day in Austin scores about 30 on the EPA's Air Quality Index (AQI)—used to measure pollution levels—and scores in Los Angeles, America's most polluted city, will at times top 100, air quality in Delhi and Beijing made our experience almost like visiting a different planet. The Environmental Protection Agency (EPA) deems scores between 200 and 300 to be "Very Unhealthy" and anything over 300 as "Hazardous."[16] Beijing's AQI has been regularly well over 100 and not uncommonly in the 300–500 range, which it was during our visit. Over ten days in Beijing, a gray smog enveloped everything morning, noon, and night. We never saw a star during the evening the entire time. Over our week in and around Delhi, temperatures were well over 100 each day and the AQI exceeded 500, incredibly ranging all the way up to 800. The air was a dingy orange, sour smelling, like burnt plastics, and penetrated our clothing and burned our eyes.

As bad as the air in Beijing was, it paled in comparison to the alarming situation in greater Delhi, home to nearly 20 million persons. Delhi's AQI has subsequently exceed 999, the AQI's maximum score, on occasion. Particulate pollution—consisting of tiny pieces of soot, metals, burnt rubber, and more that can be absorbed into tissues of the body when inhaled—is of special concern to human health in Delhi. A Harvard study recently found that 8.7 million people per year were killed by pollution from the burning of fossil fuels, the vast majority within China and India.[17] These conditions in Delhi are so bad that it was estimated that President Barack Obama's three-day visit there took six hours off his life.[18]

Other factors have aligned to grow our climate problems. The United States' exit from the Paris Agreement and the over 100 environmental policies revoked by the Trump administration—and the license they inspired other countries to take—may well compound this trend in years to come. Quick reversal of many of these decisions by the Biden administration will not heal all the damage caused. Even more concerning, many of our foreseeable

responses to climate change—from mass proliferation of air conditioning in parts of the world that have never had it before to
increasing deployment of seawater desalination to replace vanish-
ing fresh-water supplies to the construction of large infrastructure
projects such as seawalls to stave off rising seas—will all require
vastly more energy than we use today.

Consider in more detail just one of these compounding problems: air conditioning, a massive and growing source of increased
demand for electricity generated from fossil fuels worldwide. In
2000, the world consumed about 300 terawatt-hours of air conditioning, a quantity of electricity roughly equivalent to that consumed annually now by the state of California.[19] The UN IPCC
cites a study finding that as temperatures climb and incomes rise
in places like India and China, which are lifting millions out of
poverty each year, air conditioning will grow by thirteen times,
to 4,000 terawatt-hours, by 2050, and then by more than thirty
times, to 10,000 terawatt-hours, by the year 2100.[20] This is the
equivalent of adding thirty-three new Californias to world energy
consumption by the end of this century, most of which, if things
do not change, would come from fossil fuel power plants, in turn
driving ever more emissions.

It is noteworthy that this rising wave finds us in a time of singu-
lar and historic vulnerability. As a species, we have engineered the
global agricultural systems that feed us into very nearly a mono-
culture: today, about 75 percent of all food consumed around
the world comes from only twelve crops and five animal species.[21]
In terms of biomass, humans and our livestock today exceed all
wild mammals by a vast order of magnitude, our farmed poultry
account for nearly three times the biomass of all wild birds, and
plant biomass has seen a two-fold reduction in total on the planet
during the period of human civilization.[22] If any of a multitude
of potential threats led to the endangerment of the handful of
crops or species on this short list that seven billion human beings
rely upon for sustenance, the results would clearly be beyond

description. Our ability to cooperate and resolve conflicts through institutions like the UN and the World Trade Organization as means of weathering these challenges is much in question. The moment when we most need effective international cooperation to resolve complex global problems is the same when we may find it most depleted.

Understanding our situation through the lens of the carbon budget brings into sharp relief some hard truths about our current predicament. One of these truths is that, due to past emissions, many negative consequences are already "baked in" to our future. Even if we achieve the most aggressive emissions reductions targets put forward to date, temperatures will increase significantly, sea levels will rise substantially, and mass extinctions will unfold, for centuries to come. This means, of course, that conversations about reducing emissions based on longer, more realistic timetables are actually conversations about how much worse things will get and not about whether we will avoid significant consequences altogether.

No fact about climate change seems to elude people more than this one, but it is a critical one to fully bring on board. Scientists have now concluded that if people magically ceased 100 percent of all greenhouse gas emissions tomorrow, much of human life will nonetheless be transformed negatively. This reality is chillingly stated in the dry, matter-of-fact language used by the UN IPCC scientists in a 2014 report:

> Surface temperatures will remain approximately constant at elevated levels for many centuries after a complete cessation of net anthropogenic CO_2 emissions. A large fraction of anthropogenic climate change resulting from CO_2 emissions is irreversible on a multi-century to millennial timescale, except in the case of a large net removal of CO_2 from the atmosphere over a sustained period.

> Stabilization of global average surface temperature does not imply stabilization for all aspects of the climate system. Shifting

biomes, soil carbon, ice sheets, ocean temperatures and associated sea level rise all have their own intrinsic long timescales which will result in changes lasting hundreds to thousands of years after global surface temperature is stabilized.

There is high confidence that ocean acidification will increase for centuries if CO_2 emissions continue and will strongly affect marine ecosystems.

It is virtually certain that global mean sea level rise will continue for many centuries beyond 2100, with the amount of rise dependent on future emissions. The threshold for the loss of the Greenland ice sheet over a millennium or more, and an associated sea level rise of up to 7 meters, is greater than about 1°C (low confidence) but less than about 4°C (medium confidence) of global warming with respect to pre-industrial temperatures. Abrupt and irreversible ice loss from the Antarctic ice sheet is possible, but current evidence and understanding is insufficient to make a quantitative assessment.[23]

These 2014 conclusions are due to be updated by the UN IPCC in late 2021, and it seems reasonable to expect, given the failure to reduce emissions, that the new report could project even more "baked in" changes.[24]

Another uncomfortable truth is that the best efforts of the world's countries to manage emissions to date have been wholly inadequate, and this is a fairly charitable description. Commitments made by signatories to the Paris Agreement, the strongest step taken thus far to reduce emissions globally, will not even on their face achieve the agreement's stated goals of keeping temperatures below 1.5°C, much less 2°C, of warming. As of 2020, analysis indicates that even if all signatories meet the goals they set for themselves under the agreement, 2030 emissions will not fall below 2015 levels, and the most likely temperature increases we could expect would range between 2.1°C and 3.3°C by 2100.[25] Even worse, most Paris Agreement commitments only extend through 2030; the scenarios showing warming limited to just 2.1°C to 3.3°C actually assume that dramatically quicker emissions reductions will occur worldwide post-2030, after the agreement expires—an assumption for which there is scant basis today. This future, in which emissions fall all the way to zero somewhere between 2060 and 2080, is difficult to imagine coming to pass, based on today's headlines.[26]

A final truth is that it is already clear that any future strategy to effectively manage the worst of climate change cannot limit itself to simply reducing greenhouse gas emissions. Emissions must be reduced, but if we are on track to burn through the 2°C carbon budget by 2037, it is clear that we also need to invent a means of safely removing massive amounts of such gases from the atmosphere and that we must do this in ways that present no risk of unintended consequences that somehow worsen the situation. And here we find a familiar problem: the most potent methods we have developed thus far to remove carbon from the air, apart from nature-based solutions such as tree planting and preservation, also require massive amounts of energy, energy that still comes predominantly from burning fossil fuels. We are hampered at every turn by the mammoth size of our energy demands and our substantial reliance, in meeting that demand, on fossil-generation technologies to power them.

It is tempting to believe that we have not yet reached the point of no return that locks in a 2°C or greater temperature increase. In truth, however, the steady trajectory of our emissions growth over the last several decades and the political realities afflicting the leading emitting nations today strongly suggest that it is effectively behind us already—that these factors are so powerful that we are all but fated to blow right across the threshold.

CHAPTER 3

THE POWER PLANT THAT SAVED THE WORLD

How a tiny solar power plant in West Texas heralds transformation of the solar industry, the electricity business, and, perhaps, the arc of climate change.

Having gotten my start in renewable energy with West Texas wind in the 1990s, I was used to seeing wind turbines towering from mesas here and there all across the landscape. In fact, I had flown so many times from Austin to Midland over the years that I could identify many of the wind projects from the air, a growing list as the amount of wind generation built in Texas exploded from the first few hundred megawatts in the early 2000s to more than 10,000 megawatts by 2010 and nearly 20,000 today. Texas was always a "wind" state when it came to renewables, but then in the 2010s, we started to hear about the first solar projects being constructed.

As a "wind guy" my whole career up to that point, I must admit that I initially looked askance at the early Texas solar projects. Everyone knew solar was far too expensive to compete with wind, so it wasn't going to be much of a force. Slowly but surely, though, the first solar projects began popping up here and there, and then began to expand at a gallop. In 2010, there were only about 35 megawatts of solar operating in the whole state, but within just a few years there were ten times as much, and the amount has increased every year since by between 50 percent and 100 percent. Today there is vastly more new solar generation in development in Texas than wind, suggesting that while Texas may have always been a "wind" state, its future is likely much sunnier than it is windy.

Solar energy projects are generally pretty interesting to look at: sleek, futuristic, geometric, each panel somewhat reminiscent of the captivating obelisk in *2001: A Space Odyssey*. However, the Barilla solar project in Pecos County, Texas, is about as run-of-the-mill as they come. Only visible at a fair distance from a public road, the project's rectangular photovoltaic panels line up row after row, gazing upwards from a postage stamp of sand cleared out of barely there West Texas scrub trying hard to stave off the desert creeping in from New Mexico. The Barilla project is not remarkable for its looks or its size; in contrast, there are photovoltaic projects in California at least twenty times larger, staggering vistas of manmade geometric uniformity extending for what seems like forever across lunar desert expanses. Nor is it outstanding for its science-fiction appeal—that prize would surely go to one of the concentrated solar thermal projects in places like California, Arizona, or Nevada that use massive arrays of parabolic mirrors to reflect sunlight onto towering pillars hot enough to heat pure salt to nearly boiling.

Even among the solar projects within Texas, the Barilla project is not outwardly remarkable, not much different from the others constructed here and there in its general vicinity. And it will likely become even more anonymous in time, as more and more new solar projects arrive in West Texas. The fact is that today solar energy has become extremely competitive with every other power generation option in Texas and, as a result, there are more and more solar energy projects being built across the state.

Yet the Barilla project is remarkable—in fact, extremely so. What makes this project so special is the mere fact that it was built at all, given that, at the time it was constructed, no one had agreed to purchase the energy it would generate. The construction of a power plant without a prearranged purchaser for its output may not sound particularly exciting or even noteworthy at all, unless you have an understanding of how power plants are financed. What is so unique about Barilla is that, in the world of large-scale power plant financing, only a power plant expected to be competitive over its full life span on the open market gets built without a pre-agreed contract with a buyer. Heretofore, this had been the exclusive domain of certain coal and natural gas power plants. Thus, the appearance of the first solar plant built "on spec," or without a contract, signifies something completely new.

Viewed from this perspective, the Barilla project is quite different from all the other solar projects that came before it, in Texas and elsewhere. It is not overstating the point to say that the plant's status is something big, presaging a departure from solar's already phenomenal growth curve over the past ten years to an even steeper one. If Barilla means that solar power plants are now as competitive as fossil fuel power plants, its appearance points the way toward solar's ultimate success in the fight for market share in wholesale electricity markets everywhere, which in turn signals big changes for these technologies and, more importantly, for the behemoth utility companies that own them. Market-competitive solar would have the potential to remap the electric power business, in time reinventing it as no longer the most potent driver of climate change but instead as perhaps our most powerful defense against it. Plentiful, cheap, and clean power could finally decouple quality of life from the heavy carbon footprint that underwrites it today.

So, the little Barilla solar project turns out to be a pretty big deal.

■

The private roadway to the Barilla project starts at an odd spur from a feeder road off Interstate 10. The road previously dead-ended abruptly in the sand, perhaps in anticipation of a residential subdivision that never materialized. The project itself sits several hundred feet from the highway on an odd-shaped plot of land near the transmission line to which it delivers its electricity. The project's solar panels are aligned crisply across the land along mathematically perfect lines oriented to maximize energy production. Because Texas is in the northern hemisphere, all the panels are aligned along an east-west axis, angled down slightly toward the south to catch the rays arriving directly from the sun only at the equator. If you look closely enough, you can see that the rows do not have a perfect east-west orientation; rather, they slant slightly down to the east. This slant provides better exposure to the late-day western rays than to the morning eastern ones—a reflection of the fact that late-day electricity is more in demand, and therefore more valuable, than early-day electricity.

Under each of the solar panels is a set of wires collecting the generation from it, then routing the electric output to a central point, where a series of inverters transform the power generated from direct current into grid-ready alternating current. From the collection point, the output moves into a voltage transformer, which raises the voltage to transmission line level—in this case, 138,000 volts. Wires from the transformer then flow out to the point of interconnection with the grid, finally sending the electrons out to power the world.

Barilla is located just west of Fort Stockton, Texas, a small town in the middle of nowhere. Standing at the courthouse in the town square and looking west, you face 250 miles of forbidding desert, scrub, and mountains before the next outpost of civilization, which is greater El Paso. Looking north, you see more than 100 miles of the same landscape, all the way to the dry and dusty sister cities of Midland and Odessa, and to the east or south you would drive hours before reaching a town of any size at all.

Fort Stockton is the county seat of Pecos County, one of several carved from a single vast jurisdiction back when Texas joined the Union. Pecos is second in size only to its sister to the south, Brewster County, and both are massive: Pecos would fit two Delawares inside it, with room to spare, and Brewster could fit all of Connecticut and half of Rhode Island. Driving eighty miles per hour most of the way, it would take you an hour and a half to get from Sheffield, Texas, on the eastern edge of Pecos County, to Hovey Road on the western boundary.

Pecos County and Fort Stockton sit at the eastern edge of the vast, sparsely populated area between greater El Paso and the end of what Texans call "West Texas." You might think the term "West Texas" would include everything up to the New Mexico state line, given that all this area lies west of the rest of Texas, but you would be wrong. The flat lands around Fort Stockton represent the last of the Great Plains coming down from Canada, and not too far to the west of town they give way to mountainous areas like Big Bend and the Delaware and Guadalupe Mountains and the vast basins separating them. For this reason, Texans tend to see everything beyond Fort Stockton not as part of West Texas but instead as its own distinct place. Geographers agree and call this region "Far West Texas" or the "Trans-Pecos," but most Texans I know don't use either term; they just refer to the town or place they mean by name and know that out there it is east to get to West Texas and west to El Paso. The boundary between the Central and Mountain time zones also divides these regions—yes, Texas is so big that it straddles two separate time zones—which adds to the feeling of dislocation in this part of the state.

There are a number of reasons why Fort Stockton is a good place for a solar project like Barilla. First and foremost, there is plenty of sun. Fort Stockton is in the sunniest part of the state, enjoying an average of 263 sunny days and only thirteen inches of rain per year, about one third of the national average.[1] Many people think of Texas as a very sunny state, but in reality only the western quarter compares to the sunniest areas of California, Arizona, and New Mexico. Galveston and the rest of East Texas, for instance, get plenty of sunny and hot days but do not compare in terms of solar resource.

Fort Stockton also offers advantages for solar projects within Texas's unique electricity transmission system. Texas is the only state with a power grid electrically separate from that of any other state, which allows it to be exempt from federal jurisdiction. This feature is steadfastly protected by Texas's utilities and the state's Public Utility Commission, in line with the state's "less government is better" ideology. However, this independent grid, which is managed by the Electric Reliability Council of Texas (ERCOT), only covers about 80 percent of the state. The remainder is serviced by grids extending across multiple other states. ERCOT serves the biggest power markets in the state, including all of Texas's major cities and industrial customers, while the adjacent grids generally cover rural areas with fewer energy customers.

Fort Stockton sits at the sunny, westernmost edge of ERCOT, which puts it in a sweet spot of sorts: you can't get much sunnier and still be within ERCOT and its power-hungry customers. If Barilla were further west or north, it would get even more sun, but it would also mean leaving ERCOT and instead competing with the even sunnier solar projects connecting to other grids in New Mexico and Arizona for even fewer customers.

In addition to all this, Barilla fits well within the unique West Texas energy culture. This may be surprising, given the well-known conservative politics of deep-red rural Texas, but Fort Stockton has always been a frontier town, comfortable making the most of its natural resources to survive, and it has accepted solar as one of its own.

Founded in 1859 by the US Army as a frontier outpost near the clear, flowing water bubbling up from Comanche Springs, the original settlement was fought over first by settlers and Comanches and other tribes and later by Confederate and Union troops.[2] It became a permanent settlement only after the Civil War, when farmers started to exploit the springs and later the nearby Pecos River to irrigate thousands of acres of farmland in the region. The town grew as the center of a farming region, but irrigation over-whelmed the aquifer by the 1950s and the springs dried up.[3] This past is difficult to imagine today, as the absence of water is strongly felt almost all year. The Pecos River is but a fraction of its former self, its width narrow enough today to jump across at many points.

Farming abandoned Fort Stockton, but energy arrived not long afterwards to fill the void and has been there ever since. At the eastern edge of Pecos County begins the mighty Yates Oil Field, one of the largest and most prolific fields in the United States and indeed the world.

The Yates Field was discovered in the 1920s, one of several sites in West Texas's Permian Basin that energy historian Daniel Yergin calls "one of the great concentrations of oil in the world."[4] The first well drilled struck oil at just 992 feet and produced such quantities that a nearby canyon had to be turned into an impromptu storage lagoon.[5] A few years later, in 1929, another well just a few hun-dred yards away blew out and set a world production record, and in the years that followed wells were drilled all across the region.[6] Massive production from the Yates chugged on reliably year after year, decade after decade, reaching a staggering one billion barrels of oil in cumulative production in 1985.[7]

It took until the 1980s for production at the Yates to finally slow, but around then new drilling technologies such as water flooding had been developed and were applied to rejuvenate out-put. These methods were very effective and amazingly, Yates field production hit two billion barrels in 1995.[8] Starting in the late 2000s, CO_2 flooding and hydraulic fracturing technology were applied to the Yates, initiating another significant boomlet in Fort Stockton. Fracking lifted economies not just in Fort Stockton but also just about everywhere else in West Texas, only to set them back down where they started when oil prices collapsed again in 2014. Until then, oil and gas drillers and operators filled local hotels and restaurants to the brim, and nearly any Texas oil worker I run across today will tell me about Motel 6 rooms, in towns like Fort Stockton and Midland, renting for $350 per night during this period.

Yates has been one of the most prolific oil fields in the world, and it shows.[9] Drilling activity on the field's 26,000 acres has been so intensive over the past century that its pockmarked landscape is easily visible from space. Production continues today, and some see substantial life left in it. One oil executive believes that only a fraction of the five billion barrels estimated to be in the Yates have been produced.[10]

During the late 1990s, wind energy was also "discovered" in West Texas, leading to a new energy boom and frenzy—something I participated in firsthand. "Discover" is a strange term for what happened with wind in Fort Stockton and neighboring West Texas areas at this time, given how windy everyone always knew the entire region to be. Wind developers, including myself, were attracted to the area early on, but we encountered an unusual problem. While everyone knew it was windy there, a lack of public wind data from airports and the like meant that no one had information with which to prove exactly how windy it was and where exactly it was windiest. Back then, those of us looking for land windy enough for wind turbines to turn a profit had to guess about where the windiest locations were based on local lore, topography, and gut feelings. Often a landowner would explain to us: "No, it's not windy on that part of my ranch. If you are looking for the windy spot, I'll take you to a place where I don't ever stop my pickup because I can barely open the door due to the wind always blowing." Nine times out of ten what they had to tell us was better than any of our hunches.

Happily for Pecos County, the mesa formations just west of Fort Stockton were hurricane-force wind sites. When the first meteorological towers were erected to scientifically measure wind speeds in the mid-1990s, the result was, for the wind industry, like the first gusher for the Yates being drilled back in 1926. The "Great Texas Wind Rush" followed, in which wind companies competed to lease the windiest lands.[11] During the late 1990s and early 2000s, several of the largest wind projects in the world were constructed in Pecos and neighboring Upton and Crockett counties, and a steady flow has followed since. There has been so much wind generation in this part of the state, in fact, that billions of dollars of new transmission lines were added in 2013 to ferry all the wind-generated energy back to load centers in Houston, Austin/San Antonio, and Dallas. More recently, vast numbers of turbines have been constructed elsewhere in the state, including the Panhandle, the Rio Grande Valley in South Texas, and even parts of North Texas outside of the Dallas–Fort Worth metroplex.

Energy has been true to Pecos County for more than a century—first oil and gas and then wind—so when solar developers started showing up in the 2010s looking to initiate new projects, they fit right in. Energy from one source or another has driven the local economy for longer than anyone can remember, so it's not surprising that people in West Texas look past any politics around renewable energy and see solar for what it is: the latest way to build on the region's rich energy legacy.

The Barilla project was the first solar project built in Pecos County. When the project was announced in 2014, Pecos County Judge Joe Shuster summed up all these synergies nicely: "In West Texas we've got plenty of land, some with a lot of oil under it, and all of it with sunshine, which makes it perfect for solar plants like this. I'm excited to see Barilla as the first project in what I hope will soon be the 'Texas solar patch.'"[12] Like oil, gas, and wind before it, solar is now bringing new investment, jobs, and tax revenues to the region.

It is interesting to note that the electric substation connecting to the Barilla project, from which the project takes its name, was built in the 1940s, indicating that it was likely installed to serve the growing electric load required at that time by frenetic oil and gas activity at Yates field. This means that Barilla's substation and transmission line were originally built to do the opposite of what they are doing now. Instead of delivering electricity from power plants in Dallas or Central Texas to extract oil from the ground in West Texas, these lines are today delivering clean solar-generated power from West Texas back to power-hungry population centers to the east.

All this history is visible, literally, on the face of the land in Pecos County. If you fly over Fort Stockton, which many routes from points east into Los Angeles or San Diego do, you can see it all by simply looking out your plane window about thirty minutes past Dallas. Crisscrossing the landscape is nothing less than a record of the energy past, present, and future of West Texas. Vast tracts of land in all directions are pockmarked with caliche stone pads and active wells or capped drill holes, all linked by power lines, overland pipelines, and distribution lines for producing, collecting, and transporting oil and gas. Visible on the mesa outcroppings here and there are rows of dots with slowly rotating blades attached to them, each one a wind turbine churning, for the better part of each day, the ripping West Texas winds into electric power. More than a thousand wind turbines operate on the mesas in and around Pecos County. Giant transmission lines slash across the landscape occasionally, connecting the wind projects and the natural gas power plants down the road in Odessa to Dallas, Austin/San Antonio, and Houston. Last but not least, this energy collage now includes the small postage stamps of solar panels off Interstate 10 and a few other locations, with more and more of them popping up year by year.

With this final piece, the picture of Fort Stockton as a scrappy frontier town getting the most it can out of all its resources is complete. I am fortunate to have been a participant in this story over the past twenty years and have watched it unfold before my eyes. The growing wave of wind turbines and solar panels has meant changes to the landscape and so much more, including jobs and tax base, which in turn have meant improved roads, new schools, and better funding for local services. The energy wheel spins round and round in Fort Stockton, but the region has always found itself at the center, regardless of which technology reigns supreme.

■

Solar panels fit so naturally into Fort Stockton's energy landscape that you could imagine them sprouting from the West Texas earth, just like the oil and gas wells and the wind turbines seem to have before them. But of course, none of these projects was born like that. They materialized in a much less mysterious way, following the same arduous process as any other kind of privately owned, capital-intensive project. Which is to say, they were financed.

In the film *The Graduate*, when the fuddy-duddy family friend advises Dustin Hoffman's character on a career choice—"Plastics. . . . There's a great future in plastics," he famously intones beside the pool—you can see the very light go out of Hoffman's eyes. Nothing, apparently, could be more boring and soulless, or irrelevant to the things that are important in life, than the plastics business.

Many people might say the same thing about finance. What could be more boring than talking about finance, and what does it have to do with climate change, anyway? As it turns out, when it comes to power plants and our energy infrastructure generally, finance has quite a lot to do with climate change. While the subject mainly conjures images of nameless, faceless corporations in boardrooms dividing up piles of money with each other—and with their bankers and lawyers, of course—power plant finance is, in fact, about a great deal more than that. If you are interested in energy and climate change, then by the transitive property, no matter how unrelated they may seem, you should also be interested in power plant finance.

Imagine for a moment that the next generation of power plants trying to get built are instead college athletes, all waiting and hoping to be drafted into the pros for long and prosperous careers. There are many more candidates than positions, so the coaches have to pick the players whom their analysis indicates will produce the best results; the main way they decide this is by assessing how successful they will likely be in competing against other players. What has happened in the utility business for decades is that new fossil fuel power plants have always been the players selected to advance because they demonstrated the highest return on investment—a proxy for the best ability to compete in the electricity market—while renewable energy projects have generally not been chosen because they were less competitive.

This system is essentially what explains our current mix with regard to electricity generation in the United States, and most of the rest of the world as well. A great majority of the funds available to finance power plants has gone into coal, natural gas, oil, and nuclear generation and a much smaller percentage into wind and solar. In 2017, fossil fuels accounted for 63 percent of electricity generated in the United States, while just 7.5 percent came from wind and solar.[13] Worldwide, fossil fuels account for 81 percent of global energy production while less than 2 percent comes from wind and solar.[14]

The indirect result of all this, of course, is of urgent relevance to climate change. These finance decisions determine not only the technology that generates the power we require every day but also the CO_2 emissions we commit ourselves to, and they do this not just for today but for decades to come. Power plants are typically designed to operate for at least twenty years and, in fact, a great many coal and natural gas plants continue operating for more than twice that long.

From a climate change point of view, then, power plant finance is the arbiter of which power plants, and therefore what level of emissions, we bring into the world, and which plants remain on the sidelines. In its role as arbiter, the finance process essentially amounts to the foundry of our energy infrastructure—and therefore, the process to which we have de facto delegated much decision-making about our energy future, and in particular our climate future. All the work I did over the years to identify and develop new renewable energy projects was designed to position our projects to receive a favorable project finance decision, and all the work that came after—actually constructing the project and then operating it—was the result of the financing decision to invest in the projects.

Given its importance, it is key to understand exactly how this foundry of our emissions trajectory works, and how it has evolved over time. At the conceptual level, finance is simply the process by which our economic system transforms massive amounts of capital into the specific power plants that light our world. This process has been used to solve the central problem of the electric utility business since it first began. The costs of building a large utility-scale power plant are massive, but the revenues from sales of its output accrue so slowly that attractive investment returns take decades to materialize. In a nutshell, finance is the means by which the required investment is reconciled with the anticipated revenues.

In the earliest days of the utility industry, power plants were financed under a system that solved this problem in a straightforward way. From the beginning of the modern utility business in the early twentieth century through the 1980s, utility companies were able to finance their power plants under the aegis of specially created regional monopolies: the vast investments required to build them earned revenues from power sales, with prices set by the government at fixed rates extending over the life of the plant.[15] Customers might have been satisfied (or not) with the prices set, but they had no option to buy elsewhere because only the local utility was allowed to own power plants. This extraordinary system of protected monopolies and government set rates—known as the "Regulatory Compact"—served as an extremely effective means of facilitating the financing and construction of power plants, underwriting the period in which most of the country was first electrified.

As effective as this system was, it eventually gave way to change. Starting in the 1970s, policymakers in the United States and other developed countries began to gravitate toward market-oriented approaches, not only in the electricity industry but also in other highly regulated sectors like telecommunications, aviation, and natural gas.[16] Over the next few decades, federal and state governments undertook a series of reforms that diminished and, in some jurisdictions, even eliminated governmental revenue guarantees for power plants in favor of market-based approaches. These reforms have slowly but surely deregulated large parts of the industry and opened them up to varying degrees of competition, depending on the particular rules adopted by states. Today, about 85 percent of Americans are served by utilities in competitive wholesale electricity markets, and several markets have competition for retail customers as well.[17]

Not surprisingly, the change from protected monopoly to deregulation fundamentally altered the way power plants were financed. In competitive wholesale markets, companies building power plants had to come up with new ways to solve the problem of reconciling power plants' vast capital costs with the relatively paltry annual revenues. The main way this was accomplished was by creating a compelling facsimile of the governmental revenue guarantee: a corporate revenue guarantee. Instead of a sale price guarantee being established by law and regulation, it is set forth in a contract between the independent power company building the plant and the local utility company buying the power the plant will produce. These contracts are called "power purchase agreements," or PPAs.

PPAs are binding agreements between a power plant and a specific, creditworthy buyer, typically a large utility or wholesale industrial customer, under which the buyer agrees to pay a fixed price for all the electricity delivered over a fixed term, usually somewhere between ten and thirty years. An owner of a power plant with a PPA sleeps easy at night knowing that no matter what happens, its creditworthy customer is going to provide cash flow to the project for everything it can deliver to the power grid. This commitment from the buyer for all the plant's output makes raising investment in the plant easy because it resolves the most worrisome risk factor in deregulated markets: low sales prices, or finding no buyer at all at the price sought. The customer also sleeps easy at night, knowing that however deregulated market prices may fluctuate over time, the PPA ensures a fixed quantity and price of power, which it in turn can resell to its retail customers at higher prices.

In the 85 percent of the US that today has deregulated wholesale markets, the PPA has been the primary means for financing and constructing new power plants. There is another method for financing a new power plant, however, that is significantly less common: simply build it on spec, with neither a buyer prearranged nor a sale guaranteed by the government. This kind of project is financed with no purchaser commitment at all but instead with only a belief that the power it will produce will be so competitive that such a commitment is unnecessary. For revenue, such a project can look only to what it can get by selling its output into its local electrical market at whatever the prevailing price happens to be at the time it delivers.

These speculative projects are called "merchant" projects in the electricity business. The "merchant" in the name comes from the idea of a merchant selling his or her wares at whatever the going rate is among buyers and sellers on any particular day. Just like that marketplace vendor, merchant plants carry the risk of whether or not they will generate a profit, given that the price may vary over time. Merchant projects have no need of a guaranteed purchaser; indeed, they do not want a fixed sales price, because they believe they will make more selling into the open market. These projects will no doubt occasionally endure periods of low prices when power on the grid is plentiful and demand is low, but likewise, they expect to reap high market prices during periods of low supply and high demand.

The distinction between PPA and merchant power plants is a distinction between bases of project financing and something else: the scale of the opportunity for growth of each type of plant.

While PPA plants are the most common, their potential for growth is limited. In the entirety of my career in renewable energy, the main task to ensure that a project would be financed and constructed was finding a utility that was willing to enter into a PPA with my project. If that commitment materialized, most other problems the project faced could be solved. However, the requirement that a project must find a willing buyer and negotiate an often lengthy and complex PPA with them before it can proceed substantially limits the number of projects that can be built. Finding parties interested in buying, making proposals to them, and then, if the project is lucky enough to be selected, negotiating the PPA all the way to execution are time-consuming and expensive processes with uncertain outcomes each step of the way. There are limited numbers of utilities interested in executing PPAs in the first place, and there are many parties competing to secure those PPAs, which means that the PPA was the "golden ticket," and the issue was whether you were lucky enough to find it.

In contrast, merchant projects require no partner, so they can proceed anywhere that a project's economic forecasts, confirmed by due diligence, indicate it will be competitive. This universe of opportunities to build merchant plants is only limited by market conditions—if the market price seems durable enough and higher than the price the project requires to reach its target return on investment, it can secure its financing and go forward. The underlying principle is a purely capitalistic one: if the price the plant needs to reach its target return on investment can always be bid lower than the expected prevailing price, then, in a competitive market, it will always find a buyer at an acceptable price.

Heretofore, coal and natural gas power plants have been the only kind to be financed on a merchant basis, and it has been unheard of to build solar or other renewable merchant plants. This has been the case because fossil fuel power plants have produced the cheapest power, while solar plants have needed higher-than-market power prices to achieve their target investment returns. Solar projects have required an assurance of such higher prices over ten- or twenty-year periods in PPAs to be financed.

You might wonder: If solar projects require PPAs in order to be financed, but they have a hard time competing in the market on price, how have all the solar projects that have been built to date found PPA customers? Who exactly are these buyers who would choose to pay above-market prices?

The answer is somewhat complicated, but in general it is that these buyers are not actually choosing to pay more than they should; rather, they are valuing the solar power they enter into PPAs to purchase differently from the way others in the market do. Power markets are nuanced and complex, and they present different ways for customers to determine the value of power delivered from different plants. Some buyers find solar output a bargain relative to the closed universe of their own particular power price options, which may not be the same larger market into which the solar project may sell. For instance, a large power consumer like a university or a grocery-store chain might pay a higher price for electricity to its local utility than it pays if it purchases its power wholesale directly from a solar project. Other buyers might find solar the cheapest option to secure for peak energy usage periods, such as hot or cold afternoons when fossil fuel plants are at or near full capacity but energy demand surges, or they may prefer a fixed solar price to a price from a natural gas-fired plant that includes a market variable price component for its gas supply. Still others recognize value in the environmental credits that solar projects accrue via governmental or corporate programs to make the purchase pencil out as the most all-in valuable.

The upshot of all this has been that only fossil fuel power plants have been built on a merchant basis,[18] while renewable energy projects have not been because they required power purchase agreements.[19] Clearly, however, if a renewable power generation technology ever emerged with a commanding advantage versus market prices, the growth potential for the power plants using that technology, financed on a merchant basis, would be large and sustained. The scale of the opportunity to build the new technology would be essentially as large as its price advantage—which is to say, if a generating technology could undersell existing generators consistently, it would have massive potential to grow, not just supplying power to new buyers but also encouraging existing customers of more expensive plants to jump ship for their future purchases.

The unique thing about the Barilla project is that it is the first solar project to have been financed in the manner of the most dominant competitors in the utility sector, the fossil fuel merchant plants. If power plant finance is the place where the energy infrastructure component of climate change originates, the fact that with Barilla, the foundry churned out not another fossil fuel project but a merchant-solar-powered one, is monumental. For the first time, this master driver process of our disastrous climate problem has selected not the fossil fuel technology it has always heretofore tapped from among those vying for investment; it has selected, without the intervention of policy directives requiring such an outcome, a zero-emissions merchant generation technology as the most attractive recipient for power plant capital.

As detailed in chapter 5, subsequent to Barilla's construction, circumstances conspired to dramatically change the market conditions that Barilla was expected to see, and three years later, the plant's owner wrote down its investment. These circumstances are explored extensively below, revealing that the sudden and unexpected downturn in market power prices that Barilla encountered similarly surprised other new power plants, including a large natural gas power plant that was also built merchant in Texas in the same year and that filed for bankruptcy as a result. Just as the price downturn, and not the competitiveness of gas power plants generally, overturned the economics of that gas power plant, so did it overturn Barilla's balance sheet, but not the momentum behind solar power that created it. Barilla's write-down (a reduction in the estimated value of an asset) does not obscure the historic nature of its arrival.

In fact, in the annals of power plant history, the write-down puts Barilla into august company alongside another "first": the Holborn Viaduct plant built in London in 1882 using Thomas Edison's coal power generation invention. In 1880s London, streetlights were not common, and the City Corporation wanted more of them. Edison's newest invention, a coal-powered electricity generator, could supply another fresh Edison invention, incandescent bulbs, to light the streets, but the city would not contract for them unless the price was cheaper than the current option, gas lighting. To get the contract, Edison's group offered three free months of lighting and a promise to keep the price at least as cheap as gas. Edison's coal-powered generators were installed at the Holborn Viaduct, making it the first coal power plant in the world. Four years later, the plants were in financial trouble and closed.[20] Obviously, the failure of the Holborn project did not signify the end of coal power generation but rather the beginning. What was noteworthy in Barilla's case was the same as for Holborn—that it happened at all, and all that it having happened signified in the future. Barilla is in good company in trying something very different in the power business but meeting with disappointing results.

Despite the economic difficulties the project later encountered, Barilla's appearance as the first merchant solar plant suggests that a major barrier to solar infiltration into the country's power generation portfolio is falling away, a suggestion affirmed each day that passes as more solar power plants are constructed around the world and fewer coal and natural gas projects are built.

For those familiar with the dramatically falling price of solar over the past few years—a remarkable saga recounted in some detail in chapters 6 and 7—this occurrence is anything but unexpected. Close observers of the energy space over the last decade or so have thought the question was *when*, not *if*, solar would breach this wall. Relentless reductions in the costs of solar panels and associated equipment like inverters, accompanied by improvements in panel efficiency and innovations in financing techniques, have propelled solar technology to ever-cheaper output pricing.

Tiny Barilla is only one small project in a remote corner of a single market among many markets in just a single country. In the context of the world's energy infrastructure, it is practically nothing. Nonetheless, the project signifies solar energy as now a credible player in wholesale power in Texas, one of the most competitive power markets in the country. If merchant solar is emerging as a force in Texas, it is almost certainly on the same trajectory in other markets. As the broader story of advancing solar technologies makes clear, there is little reason to doubt that the dynamics that brought Barilla to fruition will in time bring many more Barillas into the world.

If Barilla signifies all these things, it also means something yet more significant: the arrival of a critical and positive development in the saga of climate change, something in very short supply these days. If, indeed, it is the harbinger of the beginning of a fundamental market change, Barilla marks the start of a dramatic transformation in how we as a species may obtain the power we need to run our modern world—from ways that drive massive CO_2 emissions year after year to ways that produce none at all. And if it does this, it will begin to help us find ways to adapt to, diminish, and conceivably one day even reverse the greenhouse gas concentrations driving the burgeoning crisis, climate change, threatening so much harm.

If Barilla does all this, it really may one day be called the power plant that saved the world.

CHAPTER 4

BARILLA'S PROGENY AND THE #UTILITYDEATHSPIRAL

The disruptive power of cheap, clean energy foretells a sea change in where our electricity comes from and who sells it to us.

arilla is a unique example of the kinds of change that cheap solar energy is bringing about, but the same forces are also rippling through our world far from West Texas. It may be easy to doubt the idea that one tiny power plant in West Texas could be so significant, but in Barilla's case, it is crystal clear that what happened with the plant goes right to the heart of vast disruptions already well under way that are having a multitude of broad consequences. In fact, change has already burst onto the scene in the electricity business—perhaps the most significant change it has experienced since its earliest days—courtesy of the same forces that brought the first merchant solar project into existence.

The largest US utilities, leveraging the many benefits of their protected regulatory status and blue-chip financial market access, have had the cheapest financing to build power plants and thus have been able to sell the most competitively priced energy. Historically, as already noted, this has overwhelmingly been electricity from coal and natural gas power plants, a fact that accounts for the carbon-heavy profile of our large utilities. For the first time ever, however, this situation is beginning to be turned on its head. In the same way that companies like Uber, Netflix, Tesla, and Amazon have disrupted other sectors of the world economy, so now are upstart companies beginning to disrupt the well-protected domain of the electricity industry, with cheap renewable energy playing the starring role in their strategies.

In fact, disruption of the utility sector has already become so worrisome that within industry circles it has acquired its own nickname: the "utility death spiral." (If you are dubious that this clunky phrase has caught on, try searching #UtilityDeathSpiral on Twitter.) The "death spiral" relates to several emergent challenges to the utility business model, but one more than any other: the rise of cheap renewable energy, primarily solar but also wind, as a means for upstart independent power companies to siphon away utility customers. Essentially, #UtilityDeathSpiral is about how our large utilities, the vast majority of which have been slow to adopt cheap renewables, are year by year losing customers to new competitors that are piling into renewables and finding ever more creative ways to bring these customers into wholesale energy markets. The theft of customers begins a fiscal hollowing-out of the financial engines that have made utilities perennial blue-chip stock companies and will in time lead to dramatic consequences if the utilities are not able to respond effectively.

As the best electricity customers—large commercial and industrial companies that use the most electricity and have the best credit ratings—are enticed to leave their utilities for cheaper power offered by upstart competitors, the utility has no choice but to raise rates for its remaining customers. This happens because the utility still must cover the relatively constant fixed costs associated with operating its fossil fuel power plants. As bills rise for remaining customers, they become even easier targets for insurgent competitors to steal. Over time, utilities become corporate receptacles of decreasingly profitable power plants that are also less and less competitive, with fewer and fewer customers to pay for them. This, in turn, makes the corporate entities owning them ever more vulnerable to what many view as inevitable government action on power plant emissions. The utilities' liabilities in time may well swamp their asset value. It is not unreasonable to imagine, in a decade or less, some of our largest utilities becoming the "Blockbuster" to the new electricity market entrants' "Netflix," or the cab company to their Lyft, if they do not change course.[1]

The dynamics driving #UtilityDeathSpiral have been brewing for years and are gaining momentum. Companies ranging from the largest corporations in the world (Apple, Microsoft, Google, Amazon, Dow Chemical, GM, Coca-Cola, Walmart, John Deere, Nestle, and McDonald's, to name a few) to the smallest (grocery store chains, local businesses, farms, and even residential developments) have been finding ever more creative ways to leave their utilities behind by securing direct access to the cheaper electricity that wind and solar offer.[2] They see benefits not only in lower prices but also in corporate stewardship and public relations, by declaring their solidarity with the majority of Americans who believe that it is important to take action on climate change—a message many incumbent utilities seem intent on not hearing, or at least not meaningfully acting on.

The steps taken by tech giants like Apple, Google, and Facebook to target sourcing 100 percent of their power needs from renewables are uniquely powerful in diminishing overall energy emissions, given that the massive server farms these companies have built nationwide over the last decade are some of the largest new electricity consumers present. Not only are these companies contracting directly with specific wind and solar projects to power these massive guzzlers, but they are also installing renewable energy technologies, like rooftop solar, within their buildings, as a means of displacing energy they would otherwise have to purchase from their local utility.[3]

Collectively, the energy consumers that are most actively abandoning utilities are their best customers, the biggest buyers of electricity with the healthiest balance sheets, and they are fleeing the utility business model in ever greater numbers. In an ominous new sign for incumbent utilities, some of this group are now encroaching directly upon the utility business by starting fledgling utilities of their own. Initially, these seem to be focusing on serving only their own energy needs, but it is not hard to imagine them broadening their approach in the future. Google, for instance, applied for and received a federal designation to sell energy directly on wholesale markets in 2010, a step both Apple and Walmart have also since taken.[4]

Carried to its logical conclusion, #UtilityDeathSpiral will result in deteriorating financial performance by utilities and other negative changes of fortune if they fail to see the signs and adapt. Utilities around the country are responding in various ways, many by deploying political and bureaucratic weapons of resistance in battles playing out in courts and obscure regulatory venues. Others—including, notably, Florida-based Nextera Energy—are taking an "If you can't beat 'em, join 'em" approach and are operationalizing renewable energy sources on a large scale themselves.

I have seen this phenomenon take shape firsthand and begin to change the industry—and have even been involved directly myself in making some of these changes happen. In 2013, one of the wind projects my company was developing became an early part of the #UtilityDeathSpiral seismic industry shift. The Logan's Gap Wind Farm, located in Comanche County, Texas, was a project that my company had been developing for a couple of years when we started trying to find buyers for the project's energy. We began talking with the usual suspects—the two largest utilities in Texas at that time, TXU Energy and NRG—but each faced such serious financial issues then that they were not creditworthy enough to back large power purchase contracts. As a result, we broadened our search and eventually reached out to a buyer who had never bought renewable energy directly from a project before: Walmart. The company owned stores throughout Texas that consumed so much electricity that many years before they had formed their own wholesale power company to supply them. Normally, Walmart's power company bought the power they needed from other utilities, but we approached them about buying directly from our project instead and offered them a price and contract structure that got their attention. It took us nearly a year to complete the deal in a way that met all of Walmart's requirements, but in the end the agreement was signed. We later sold the project to a larger company, Pattern Energy in San Francisco, which still owns it today.

At the Logan's Gap project commissioning, Walmart's vice president of energy, Mark Vanderhelm, explained why they decided to go forward with the deal: "Walmart has a goal to be supplied by 100 percent renewable energy, and sourcing from wind energy projects—like the Logan's Gap Wind facility—is a component in the mix."[5] Since then, Walmart has proceeded to sign agreements with many more wind and solar projects in their effort to increase their use of renewables, leaving their supply relationships with traditional utilities behind to do so.[6] Walmart has a very public commitment to becoming 100 percent renewable in the future, but I can say with firsthand knowledge that the culture of cost-cutting at Walmart extended all the way through the organization to the individuals with whom we were negotiating regarding the Logan's Gap project. They loved that it was a wind project, but the low price we had to offer was even more important for the deal to have been reached.

The disruptive forces changing the way we get our energy today are, to a great extent, making energy cleaner primarily by using the laws of economics rather than "green" mandates issued by governments. Renewables are major drivers of these changes, but change is also occurring within the fossil fuel-generation world. Today's historically low natural gas prices are allowing natural gas power plants to steal market share from coal plants, even as coal is facing escalating environmental-compliance costs. Many see this as a double blow from which the coal industry is unlikely to ever recover, at least in the United States. This switch from coal to natural gas is reducing emissions substantially: natural gas combustion produces only about 50 percent of the greenhouse emissions that coal does. Coal's problems will likely worsen. Since 2012, persistently low natural gas prices have been followed by record-setting new-build announcements for gas generation in 2017 and 2018, alongside large new wind and solar project announcements. The scale of this new gas build is massive: forecasts call for thirty-six gigawatts of new natural gas generation to be built over these two years, equivalent to about half of Texas's total generation being added to the existing US generation base.[7] In 2019, US natural gas generation increased by 8 percent.[8]

Given the historical and present dominance of fossil fuels in meeting our power needs, and these recent massive forecasts for new growth in natural gas generation, it is heady stuff to assert that renewables (and solar in particular) are poised to meaningfully change this balance. Put differently, if Barilla is going to save the world, it has an awful lot of work to do.

If Barilla signifies a shift in which solar eventually becomes as dominant as fossil fuels are today, it is useful to recall the long road that coal power took before it reached the pinnacle. Coal's rise required massive associated infrastructure, first in cities but eventually almost everywhere, along with a completely new regulatory structure protecting financing of the expensive systems. It was not until the 1920s that most major cities were electrified, and not until the 1950s that most of rural America joined. Given the accelerating pace of modern change and the mounting economic advantages of solar and other renewables, it seems likely that these changes can occur much more quickly today than in coal's time, although certainly not overnight.

But there can be no doubt that it is coming. While little about the coal or natural gas power industries suggests that their pricing can get cheaper—and several factors support the idea that each will more likely become more expensive, including, notably, the growing demand for gas to supply all the new gas power plants— there is every reason to believe that prices of solar panels are going to continue to fall. After all, plummeting solar prices have been based on a consistent and accelerating wave of technological, man- ufacturing, construction, and financing innovations over the past two decades that shows no signs of abating.

For many who care passionately about climate change, it may be appealing to believe that fighting for new laws simply requir- ing the use of renewable energy is the most direct and effective way to address the climate crisis. Indeed, much new legislation is necessary, and one can expect to see action on this front now from President Biden, who famously claimed in the final 2020 presi- dential debate, "I would transition from the oil industry, yes. . . . It has to be replaced by renewable energy over time." However, acknowledging the political dysfunction regarding what to do about climate change that today afflicts not only US leaders but also those in many other industrialized countries, the benefits of economics-centric approaches come strongly to the fore.[9] In this light, one must acknowledge that, if a way emerged to cause elec- tricity market players of all stripes to start working, enthusiastically and without delay, to turn their hundreds of billions of dollars in corporate resources into renewable power plants instead of fossil fuel plants—all as a calculation of pure self-interest, rather than litigable government dictate—then a vastly more direct, rapid, effective, and potent means of improving the climate would be at hand. And this is exactly the kind of alternative that Barilla signifies.

Barilla announces a fundamental change afoot in the electric- ity industry, a change that is not a prisoner to political debates, elections, or the whimsy and vitriol of the legislative and regula- tory processes, but that instead is driven by the same engine by which the electricity industry already lives and dies: the engine of competitive industries dueling in markets to supply the cheapest energy.

From a broader perspective, the phenomenon that Barilla represents is the chance to reverse this powerful force and make it work for us instead of against us, offering the possibility that we may, for the first time ever, meaningfully press capitalism into the service of confronting climate change, when all it has ever done before is give birth to and nurture it. It was the alignment of capitalism's most powerful driver—economic self-interest—with energy exploitation that has enabled us to begin to change earth systems on a planetary scale. What Barilla represents today is the possibility of aligning these same powerful forces in the service of mitigating such changes.

CHAPTER 5

HERE TO STAY

INCONVENIENT TRUTHS FOR
RENEWABLES OPPONENTS

*Addressing likely arguments against the idea that
the Solar Age has begun.*

s someone who has worked for the past twenty years trying
to get renewable energy projects built around the country, I
expect that some will view the thesis put forward in the last
chapter about the dawning dominance of solar energy with skepti-
cism, if not disbelief. From my years working to secure support for
wind and solar projects from Oregon to Alabama and from Nevada
to New Hampshire, I know firsthand that there is no shortage of
renewables skeptics out there. I generally divide these opponents
into two groups: those with genuine and legitimate objections to a
particular project's environmental, viewshed, or other impacts, and
those with a deep-seated, philosophical opposition to the whole idea
of renewable energy. My team would present information in public
hearings and community meetings to respond to the concerns of
both groups and, while we were sometimes able to eventually find
support among the former, nothing ever satisfied the latter.

A favorite tactic of the hard-core philosophical opponents
of wind farms is to accuse proponents of trying to bring "wind

turbine syndrome" to their community. The theory behind this "syndrome" is that wind turbines emit infrasound—low-frequency sound waves too low to be heard by human ears—that causes a list of ailments to people living in proximity to turbines, from headaches to panic attacks to nausea.[1] These symptoms, critics maintain, are only the beginning; spread across a community, they would cause the wind farm's neighbors to need to relocate, resulting in depressed property values and financial losses for everyone. I heard firsthand these charges leveled against wind farms we tried to develop in places like Pennsylvania, Maryland, and Alabama.

The problem with this theory—which has gone viral online among anti-wind groups (try Googling "wind turbine syndrome" and note Donald Trump's claims that wind turbines "cause cancer")—is that the facts strongly suggest it is not true. Virtually anyone living in an urban area, for instance, is bombarded by infrasound every day but suffers no such ailments.[2] Moreover, peer-reviewed research examining all published studies on the subject concluded that there is no "direct causal link between people living in proximity to modern wind turbines, the noise they emit and resulting physiological health effects. If anything, reported health effects are likely attributed to a number of environmental stressors that result in an annoyed/stressed state in a segment of the population."[3] Despite this kind of categoric objective analysis, true wind "haters" continue to assert that the syndrome exists. And while "solar panel syndrome" only raises a few hits in a web search, no one should be surprised to know that similarly unfounded bases for opposing solar projects are increasing.

The fact is that even if we are on the verge of a historic shift into the Solar Age and a new moment in our energy history, it still may not be easy for all to see it. Historical events of consequence do not generally arrive neatly packaged, after all. More often than not, they are just a single moment plucked from a multitude in flow—each one nuanced, complex, deeply interrelated to those that come before and after it, and open to some degree of interpretation. Take, for example, the commencement of the second energy moment discussed in chapter 9. It was not likely obvious to anyone in 1712 that Thomas Newcomen's invention, the steam engine, would go on to change the world but, over time, this event revealed itself as the start of something entirely new in ways that it likely took decades to begin to appreciate.

The appearance of the Barilla project is no exception to this aspect of history. I have put forward the argument that it represents much more than it seems—nothing less than the dawn of a new age of clean energy—but this judgment will of course only be confirmed, or not, by historians at some point far in the future. In the meantime, it is fair to acknowledge that in our deeply polarized country today, essential facts concerning the project and my arguments about it may be read one way and another, facts that could temper the breadth of what we may reasonably read into its construction and the scale of change it represents. In this regard, it is appropriate to address head-on the ways in which opponents of clean energy will inevitably argue that the conclusions drawn here about the market competitiveness of Barilla and its progeny, and what it all represents, are off the mark. A project like Barilla couldn't signify much of anything with staying power, these people may say, because renewable energy is a mirage, a fraud, or an imposter, for one reason or another.

When it comes to anti-renewables claims, the granddaddy of all arguments looms largest among the naysayers: that it is propped up either mostly or entirely by government largesse and policy, not real economics. There is a bedrock belief among renewable energy opponents that no wind or solar project could ever be considered "market competitive" due to the fact that it receives government subsidies. This is both the easiest argument to anticipate—in the twenty years I have worked in the renewable energy industry, it is the charge I have heard most often from opponents to our projects—and, happily, the easiest to refute, provided one has the patience to wade through a bit of history.

It is true that solar and other renewable energy projects receive subsidies, both in the form of federal investment tax credits and, often, state and local economic development incentives. It is also true that these subsidies are significant, amounting to a substantial portion of a project's capital costs. What many critics of solar and other renewable energy subsidies often forget, however, is that it is likewise true that all their competitors—coal, natural gas, and nuclear generation—also benefit from a plethora of federal, state, and local tax subsidies and incentives, and they have for a long, long time—much longer than solar energy has.

Many Americans do not realize that the central issue in their energy source selection is not choosing *between* subsidized and unsubsidized energy, it is choosing *among* subsidized energy sources. Subsidization across energy technologies is varied, substantial, and deeply pervasive. Every generation of technology providing any significant portion of the country's energy needs has been the beneficiary of substantial and critical government support. Any honest evaluation of these programs would also conclude that this support has paid for itself many times over in terms of the phenomenally resilient energy infrastructure that our country enjoys today and has for nearly a century now.[4]

Those who believe wind and solar are largely or entirely propped up by government subsidies point to the significant tax incentives provided to the industry today. The same people often believe that wind and solar subsidies vastly exceed those received by other energy sources, particularly fossil fuels—or even that fossil fuels have not received subsidies themselves. I wrote an op-ed in 2017 in the *Houston Chronicle* that mentioned the idea that wind energy subsidies were passed in order to "level the playing field" between fossil and renewable energy sources, and some of the public comments in response to the article summed up opposing views about renewable energy subsidies very succinctly:

> If it needs to be subsidized then it's just not working for me. Oil doesn't need my money, in fact my roads are paid for by my gas

tax. When wind, solar and/or any other green energy can make it on its own, I'm all for it. Otherwise you are just trying to get me to fund your retirement account. —Eugene

How creative—now "level the playing field" is the new description for government subsidies—which we can't afford and now thanks to fracking don't need. —Virginia[5]

The gist of Eugene's and Virginia's complaints is very clear: (1) renewables get subsidies; (2) subsidies are a waste of taxpayer dollars and distort markets; and (3) we don't need renewable energy because without its subsidies it could not compete with fossil fuels, which are not subsidized.

To get a sense of where the objective support behind Eugene and Virginia's viewpoints come from, consider the table below. It comes from one of the few high-quality, exhaustively researched, and, most importantly, unbiased studies on the topic of comparative energy subsidization that can be found. Bearing the exciting title "Energy Tax Incentives: Measuring Value Across Different Types of Energy Resources," the study was completed by the nonpartisan Congressional Research Service (CRS) in 2015, updating a prior version published in 2011. (For anyone trying to keep score, both publications occurred when one or both houses of Congress, to which the CRS reports, was controlled by Republicans.)

The graph shown here, reproduced from the study, reports cumulative tax incentives provided to fossil fuels, renewable energy, and programs to improve energy efficiency over a 41-year period, from 1977 to 2018.[6] When you look at the graph, first place your hand over the left side, up to the year 2009, and consider only the information visible for the years 2010–2018. This part of the graph is mainly what is seen by those like Eugene and Virginia, who believe renewables are overly subsidized. Between 2010 and 2018, wind and solar are shown to have received approximately $72 billion in subsidies, while fossil fuels received approximately $42 billion. This gap of about $30 billion does indeed indicate a

(FY1977–FY2015)

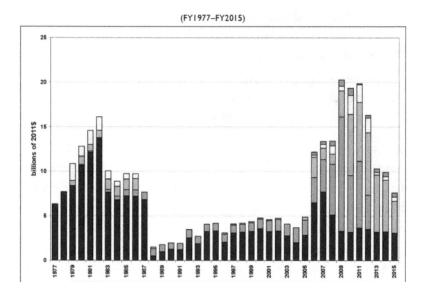

Source: CRS using data from the Joint Committee on Taxation and Office of Management and Budget.

large advantage for renewable energy relative to the other sources, one large enough to seem capable of skewing wholesale markets. If one considered only this information, it would be apparent that subsidies to renewables have vastly outpaced subsidies to other energy sources.

Next, remove your hand to expose the full graph. The left end of the graph reveals the full extent of tax incentive energy subsidies provided by the federal government since 1978, and it depicts a dramatically different story. From 1978 to 2010, renewables received approximately $25 billion in subsidies, while fossil fuels received about $150 billion.

The full graph reveals that in total, over the forty years between 1978 and 2018, renewables received approximately $97 billion in subsidies, scarcely half of the $192 billion provided to fossil fuels. Logically, if you were the kind of person who believed the subsidies given to renewable energy between 2010 and 2018 were substantial enough to sway wholesale power prices to their benefit over that term, you would also have to acknowledge that giving fossil fuels nearly twice that amount over a much longer period had to have affected these markets even more substantially.

Thus, Eugene and Virginia's arguments are supported only if you cherry-pick certain facts; looking at the full picture reveals their conclusions to be erroneous. And, while renewables have been significantly subsidized *recently*, the further back in time you go, the greater the percentage of cumulative subsidies going to the fossil fuel power sector emerges.

Most studies do not go back any further than 1978 because that period represents the introduction of subsidies for renewable energy, but in fact federal and state policies to subsidize the oil, gas, nuclear, and coal industries predate that year by more than a century. Governments have subsidized coal since the nineteenth century, oil and gas since the beginning of the twentieth century, and nuclear energy since the 1940s, with a large fraction of these subsidies going ultimately to electricity provision and the remainder to transportation fuels.

Given the preoccupation of renewable energy opponents with the myth of a bias favoring subsidies for renewables, consider the following conclusion from an article summarizing one of the most comprehensive efforts to date to estimate the cumulative total of fossil fuel subsidies, published in 2006 by the law journal at Loyola University Chicago:

> Since the inception of the percentage depletion allowance and the [Intangible Drilling Costs] deduction [in 1926 and 1916, respectively], the United States has spent between $370 and $391 billion in tax subsidies for fossil fuels, *an average expenditure of approximately $4.5 billion every year for the last eighty-seven years* [emphasis added]. Moreover, these amounts represent the tax expenditure figure only, and do not include subsidies that directly and indirectly benefit the oil and gas industry or other externalities that are more difficult to measure. Taxpayers also support government subsidies for transportation infrastructure, energy security costs, research and development subsidies, and Strategic Petroleum Reserve maintenance costs. Furthermore, these figures do not take into account externalities that flow from fossil fuel use, such

as localized pollution, agricultural crop losses and loss of visibility, planet-wide environmental costs such as global warming, water pollution costs such as oil spills, noise pollution, the environmental impact of sprawl, and travel delays and subsidized parking, all of which cost Americans both money and quality of life.[7]

These findings suggest that fossil fuel subsidies, when considered further back than 2010, dwarf those provided to renewables by a truly massive order of magnitude. This advantage, of course, would be much greater if the lens were widened to include subsidies beyond tax incentives.

It is easy to understand how people develop strong, and wrong, opinions about energy subsidies. Politicization around the question of energy subsidies is so profound that it is difficult to find unbiased information like that provided in the CRS study cited above. Just about everyone with a dog in the hunt takes the opportunity to spin the data on this question to their advantage. Consider the following examples:

In 2011, the Nuclear Energy Institute commissioned a study comparing energy subsidies over the period of 1950–2010, which, unsurprisingly, concluded that nuclear subsidies were smaller than those of other technologies. "The common perception that federal energy incentives have favored nuclear energy at the expense of renewables, such as wind and solar, is not supported by the findings of this study. The largest beneficiaries of federal energy incentives have been oil and gas, receiving more than half of all incentives provided since 1950."[8]

In 2013, the World Coal Association trumpeted the same 2011 Nuclear Energy Institute–commissioned study but used it to claim that "contrary to what is frequently stated and widely believed, federal incentives and support for renewables are much larger than those for fossil energy, and . . . this imbalance is increasing."[9]

In 2011, DBL Investors, an investment fund focusing on sustainable strategies like renewable energy, published a report comprehensively tracing US energy subsidies from the nineteenth century through today. This study focused primarily on subsidies deployed during the first fifteen years of each subsidy's lifespan, arguing that subsidies in this time period were crucial in determining later commercial viability. It concluded that "the federal commitment to [fossil fuels] was five times greater than the federal commitment to renewables during the first 15 years of each subsidy's life, and it was more than 10 times greater for nuclear."[10]

In 2017, the American Wind Energy Association (AWEA) released an analysis that showed wind to be massively under-subsidized relative to fossil and nuclear. Claiming to have completed "the most comprehensive review of energy incentives to date," AWEA concluded that, between 1947 and 2015, fossil energy received $619 billion in subsidies, nuclear $197 billion, wind $27 billion, and other renewables (including biofuels) $114 billion. Cumulatively, AWEA put fossil fuels at 65 percent of all subsidies, nuclear at 21 percent, and renewables at 15 percent.[11]

During his confirmation hearing for the position of Secretary of State, former Exxon CEO Rex Tillerson was asked about subsidies received by the company. He responded, "I'm not aware of anything the fossil-fuel industry gets that I would characterize as a subsidy. Rather it's simply the application of the tax code broadly, the tax code that broadly applies to all industry."[12]

Clearly, there has been a great deal of spin regarding energy subsidies, but the bottom line is that solar and wind energy is no different in receiving significant public subsidies from every other major electricity generation technology. Notably, these subsidies do not make renewables uncompetitive any more than they make oil, natural gas, or nuclear power uncompetitive. The fact is that when oil, gas, and nuclear generation compete in the market today, they stand on the shoulders of the massive subsidies they received in the past, subsidies that matured their technologies, supply chains, capital structures, and market delivery mechanisms and that made their final product better and cheaper. In this sense, the subsidies that the federal government has provided to all major energy sources—fossil fuel, renewable, and other generation industries—may be better thought of as smart national investments we have made as a country that have borne fruit in the form of the extraordinarily robust and dependable electricity infrastructure of the United States.

To grasp how subsidies can advance the competitiveness of power technologies like wind and solar, it may help to look at an earlier example where they were successful. In 1980, the federal government created a tax incentive for the natural gas industry called a "production tax credit" to promote a new, unconventional oil and gas drilling technology that was then in its infancy. The government had already provided grants to fund test wells drilled using this technique, developed as part of the evolving response to the 1970s energy crisis with the aim of increasing the domestic energy supply. This drilling technology was promising but risky, so Congress passed the tax credit (known as "Section 29" credits) to incentivize investors to put capital into it. Unlike a direct grant from the government, production tax credits worked by giving a well owner a tax credit for each thousand cubic feet of natural gas produced, meaning that the credit had to be earned not just by investing, but by investing in wells that used the new technology to actually produce gas. With the credits in place, the technology was able to draw investors and was subsequently refined and improved sufficiently, over the next two decades, to become economically viable on its own.

This unconventional drilling technology was horizontal drilling and hydraulic fracturing, often known today as "fracking." Research by the prestigious American Energy Innovation Council (AEIC) examined the history of federal support for fracking technology in a lengthy report, which estimates that between $10 billion and $20 billion (in 2011 dollars) was provided to nonconventional gas producers over the years 1985 to 1991.[13] There is no reason to suspect bias in this study; the AEIC is about as objective an information source as you can find, with a board that includes CEOs of large coal-intensive utilities like Southern Company, Pacific Gas & Electric, and Dominion. In assessing the overall effectiveness of the subsidy program, the report concluded:

> At recent rates of production, shale gas production alone is roughly estimated to bring on the order of $100 billion of gains to consumers each year. Government expenditures on unconventional gas R&D between 1976 and 1992 are estimated at $220 million ($473 million in 2011 dollars). Although total Section 29 tax credit expenditures associated with unconventional gas have not been quantified, Section 29 credits for unconventional gas have been estimated at $6 billion ($10 billion in 2011 dollars) between 1985 and 1991, and other reports have estimated total expenditures perhaps twice that amount. Even presuming tax credits cost substantially more than these estimates, the economic benefits of unconventional gas have repaid many times over the public investments that enabled and accelerated unconventional gas technological development.[14]

Few Americans would take issue with the idea that the subsidies provided to the natural gas industry, which supported the creation of fracking, did not constitute an effective investment in terms of achieving its objectives of enabling vastly greater domestic oil and gas production and reducing our reliance on energy imports from countries hostile to us.

It is not a coincidence that the Section 29 production tax credits passed to support the development of fracking bear a striking resemblance to the Section 45 production tax credits later enacted by Congress to support wind energy. These wind tax credits were signed into law by President George H. W. Bush in 1992 and have been extended by every administration since, Republican and

Democrat, even including the Trump administration. In 2005, Congress tweaked the wind tax credit structure to create an investment tax credit for yet another promising technology, solar energy.

It seems to be a product of today's political climate that subsidies are reviled as government handouts rather than viewed as investments made by government to secure important policy goals. The fact that the United States has massively subsidized energy resources in significant ways on a more or less ongoing basis since the nineteenth century should not be surprising. Ensuring ample access to fundamental goods and services like energy resources and electricity is, after all, one of government's essential jobs. Access to energy is routinely cited as a matter of national security and has often played a role in decisions to commit military troops. Most observers would enthusiastically agree that the United States has spectacularly achieved this end by consistently subsidizing the energy sector from an "all of the above" perspective, and the result has been the world's preeminent energy infrastructure system, more robust than any other on the planet. Government subsidies are prudent and customary when used to support new technologies to maturation and market deployment, and holding renewable energy technologies to a different standard than fossil fuel technologies is both shortsighted and disingenuous. We have much to show for our investments in energy, and perhaps we are taking it all for granted as these investments become politicized.

So, is the Barilla project a "fake" because solar energy receives governmental subsidies? It seems clear that in being subsidized, solar is on level ground with other major power generation technologies. Fossil fuels are dominant today in substantial part because they stand on the shoulders of all the subsidies they received over the decades, in the same way renewables may someday soon. Given the total numbers, renewables appear to have been, on the whole, less subsidized than their fossil and nuclear competitors, but, as stated above, drawing too fine a line is not particularly helpful given the pervasiveness of the practice. Ultimately, on any given day, the market is where we find it, and today subsidized solar has for the first time competed successfully with subsidized natural gas, wind, coal, and other generation—making it no less "real" a merchant business case than any other.

There are other reasons naysayers may take issue with the idea that Barilla is a big deal. Maybe some will argue it is a project that works but could only work in a place like Texas. Barilla's location in sunny West Texas does provide it with a superb solar resource, the intensity of which can be equaled in only a few other areas of the United States. The Texas electricity market is also uniquely deregulated and therefore more open to newcomers, and functional in many ways that other state and regional markets are not. Of all the grids I have worked in developing new projects around the country, Texas's main grid is my favorite because of its fair and transparent rules and administration; many other grids are run in ways that benefit the local utilities and the largest power conglomerates. Did the first solar project to be built "merchant" reach this critical competitive juncture only because of the extraordinary solar resource in West Texas? If so, is it effectively a one-off phenomenon, not replicable in any of the many less sunny markets in the rest of the country? Or does merchant solar only work because of some unique regulatory feature of Texas's market design?

There is logic to both arguments, but they are ultimately not persuasive. While West Texas is sunnier than most of the United States, it is also outranked in this regard by a surprising number of states. In terms of utility-scale solar capacity factor, Texas comes behind Arizona, California, Colorado, Idaho, Kansas, Nevada, New Mexico, Oklahoma, Oregon, Utah, and Wyoming.[15] Also, the state's best resource areas are located within a fairly small part of West Texas, meaning that much of the rest of the state demonstrates solar potential closer to that of Louisiana and Arkansas than to that of New Mexico. In addition, wholesale energy in Texas is priced much lower than in most other US energy markets. This means that if a project like Barilla is competitive in

ERCOT, where average wholesale power prices are in the $25 per megawatt-hour range, then the higher wholesale prices in other areas would make less energetic solar projects in these areas competitive. Moderately sunny Upstate New York, for instance, may be competitive for solar projects given New York's $55 average price of energy per megawatt-hour.

On the regulatory question, because federal law requires each US grid to be open to competition in wholesale energy markets, Texas's particular rules governing how power plants compete are not ultimately much different from those of most other states. While it is true that certain markets, like ERCOT, are more open than others for obtaining the approvals and rights necessary to build a merchant plant, these are questions of degree, not binary matters. Other helpful aspects of ERCOT's inner workings do make it a favorable grid for pursuing new projects, but ultimately one that is not terribly different from others across the country. For instance, although ERCOT is not regulated by the federal government due to the fact that ERCOT's solitary focus on Texas means it does not engage in interstate commerce, ERCOT still conforms to virtually all Federal Energy Regulatory Commission requirements governing other US grids.

Perhaps the biggest argument against the idea that Texas has any large advantage when it comes to building solar merchant projects is the fact that Texas hasn't even been a leading state when it comes to solar power generally. If Barilla owed its success as a merchant plant to Texas's favorable solar resource and market rules, you would expect the state also to be a leader in total installed solar power. In fact, this was far from the case in the year Barilla was constructed; Texas did not even rank in the top five states for installed solar projects, sitting behind sunnier states like California, Arizona, and Nevada but also North Carolina and New Jersey, which each have far inferior solar resource and generally more restrictive market rules.[16]

Naysayers may also argue that Barilla isn't worth paying much attention to because it was built to check a corporate box, not to make a profit. If its location doesn't make Barilla a one-off anomaly, then perhaps its reason for being does. Bearing in mind that the company that built Barilla, First Solar, is a major, publicly-traded international solar company with a vertically integrated structure—meaning that the company doesn't just build solar projects, it also manufactures the solar panels used in those projects—then perhaps the company built the project not for its merchant merits but because it would provide some special synergies with the deployment of its own equipment into the field. Could it have been these synergies, or some other corporate benefit of building the plant, that served to enhance its balance sheet or increase its stock price, and that thus drove the decision to approve it, rather than the belief that the project would profitably compete on its own?

This is definitely a question worth asking. First Solar had about $2 billion in cash and securities on hand in 2014, a vast sum relative to the estimated $27 million it cost to build Barilla.[17] Might such a large company have cared less about making a profit on a relatively small expenditure like Barilla than it did about another aspect of its value to the enterprise?

It can be difficult to understand a corporation's rationale with regard to a given undertaking because board decisions are often buried in the corporate minutes and hidden away from the outside world. Fortunately, however, this is not the case with Barilla, as First Solar has been fairly open about its reasoning. In an interview published shortly after the project came online, Colin Meehan, the company's Texas representative at the time, shared the official rationale behind the decision to build Barilla:

> Our thinking was twofold with Barilla. First, the cost of solar has dropped so dramatically. It is now competitive with new and sometimes existing generation—certainly peaking generation. The second component of our thinking being that the Texas market is one of the greatest potential solar markets in the country, if not the world. Texas has tremendous solar resource,

significant industrial and residential load, and the third part is
they do have a competitive market. The structure of the Texas
market is essentially that if you build it and can compete in our
market, bring it on. That is the perspective that we took. . . .
Here in Texas, we've got the ability to essentially build to the
market. Our view is that the market for solar in Texas is going
to be strong for many years to come.[18]

Meehan's explanation was confirmed by Alexander Bradley,
First Solar's CFO, during a 2017 stock analyst call: "The
30-megawatt Barilla project was originally developed to sell power
in Texas on an uncontracted basis in order to help penetrate the
Texas market, as well as to provide a test site for the implementation of new technologies."[19] It is clear from these statements that,
while First Solar did expect to reap multiple benefits by building
Barilla, the company built the plant first and foremost because
they believed it would be profitable as a merchant plant on the
open market.

These statements align with others made about the project
around the same time, both by representatives of First Solar and
industry observers. In First Solar's annual 10K filing for 2013, it
twice disclosed to investors that it was building Barilla as a merchant power plant.[20] Trade publications anxiously awaiting completion of the project due to its unique merchant structure also
remarked on it. As one article published in *Global Corporate
Xpansion* magazine put it: "One of the industry's most closely
watched projects is in Pecos County, Texas, at First Solar's 22 MW
Barilla Solar Project, which is expected to begin commercial operation later this summer. Developers are very interested since First
Solar will . . . offer the output to customers in the deregulated market . . . on a speculative basis, without first locking in a long-term
buyer for the electricity."[21]

It probably would not be fair to say that building the project
for the sake of its own financial returns was the sole reason for
pursuing it. By building Barilla as a merchant plant and having it

succeed, the company would have not only a profitable asset but also a roadmap for building many more projects without a power purchase agreement. Although Barilla was financed on First Solar's balance sheet, a successful merchant project would prove to the world and, more importantly, to lenders and tax investors in the project finance community, that solar projects could be financed as merchant plants just as many fossil fuel plants are— something that would surely be helpful to the company's broader growth aspirations.

If First Solar built Barilla not only on the project's merits but also to enhance its own opportunities in selling solar technologies, does this diminish the case for inferring, from its arrival, the kind of hard math that confirms solar's emerging market competitiveness? No; in fact, the opposite is true. If by building Barilla First Solar also hoped to prove that they could build many more Barillas, their decision to go forward means they were doubling down on the belief that the economics are solid.

Even if you were inclined to read this ulterior motive as having tipped the balance in First Solar's decision to go forward with the project, you would have to acknowledge that, at worst, the decision had to have been a close one in the first place. As with any other company, First Solar's board of directors owes a duty to the company's shareholders to maximize profits, meaning that it is not in the business of building power plants that will lose money or of unduly risking investments by putting them in precarious financial situations. The business case for building the project on a merchant basis had to have included figures very near to the company's standard investment return thresholds for it even to be considered for approval. Ultimately, it seems likely that these other factors only served to accelerate, by a short time, the appearance of a plant like Barilla.

■

Another charge that could be leveled against Barilla is that, if it isn't a one-off anomaly for one reason or another, it is not really a merchant plant. This argument can be made because First Solar has acknowledged that, since some time after the plant came online, it has operated Barilla somewhat differently than it would a purely merchant plant. For example, instead of strictly selling its power into the grid as it is produced, the company has on occasion entered into power sales contracts. Privately owned power plants do not often share how they dispose of the power they generate, but we are again fortunate that First Solar has made public statements on this point. At times, the company has declined to describe Barilla as purely "merchant," instead calling it an "open contract plant"—a categorization that First Solar's Meehan has described as being "a distinction without a difference" when it comes to the plant's operations strategy. He goes on to explain:

> I think the typical interpretation of a merchant plant is one that is strictly selling into the real-time power market, for example in ERCOT. . . . We have a variety and mix of different contracts with off-takers [for Barilla]. The one we publicized the most is Rice University, where we sell a portion of Barilla output for the next couple of years. We don't want to restrict ourselves to only one approach—a strict merchant approach. Nor are we really in a place of looking for a long-term, twenty-year PPA for that site. We built it. Now we want to see what different kinds of contracts and hedges we can build around it, and how we can make money. It's kind of a mix.[22]

So rather than selling all its output to the power market at whatever price is then prevailing, as a purely merchant plant would, the plant has sold some portion of that power under contracts with set prices for fixed periods of time, mostly of a fairly short-term nature.

On the question of whether this new information changes Barilla's status as a true merchant project, it does indeed amount to a "distinction without a difference," for at least two reasons. First, the contracts that Meehan describes were executed at pricing levels competitive with prevailing market prices. Regarding the Rice

University contract that Meehan references, Richard Johnson, director of Rice's Administrative Center for Sustainability and Energy Management, confirmed that the university was able to secure the same pricing from Barilla that they would pay for electricity from fossil fuel generation. "We were able to specifically procure renewable energy—in this case, electricity generated from solar arrays in West Texas—with no increase in cost," Johnson said after signing the contract.[23] MP2 Energy, the company that arranged the transaction between Rice and Barilla, made this same point in its statements at the time. "This is a game changer for solar energy. We can now offer solar energy at the same rates as traditional gas or coal," said MP2's CEO Jeff Starcher. "It's a lot easier for a business to commit to true renewable energy if we can provide a reasonable offtake term and a competitive price."[24] So if the power sales contracts that Barilla has entered into have come with prices that are attractive relative to natural gas and coal prices, it seems illogical to argue that the contracts prove the project is not competitive.

The second reason why power sales from Barilla under a contract do not change the historic nature of the plant's construction is that the plant's operations after construction are irrelevant to the question of whether it was financed as a merchant solar project. The basis on which it was approved for construction is the key factor, not whatever might have come later. What is critical to understand about Barilla is that, for the first time, a plant with solar-generated electricity was deemed capable of competing on price in the market if constructed, in the same corporate crucible in which it was determined that fossil fuel-plant energy met that requirement. The bottom line is that Barilla had no contract for sale of its output in place when the project was approved, and this is the barrier that Barilla broke through. There may have been hopes that other arrangements would materialize—arrangements like the Rice University contract—but the central fact was that First Solar was confident enough in the value of the project to build it despite the absence of a power sales contract when it was approved.

The final argument against the claim that Barilla represents a sea change in solar's competitiveness is, on its face, likely the most compelling: a few years after start of operations, the project had severe financial problems. It is a hard truth that Barilla did not earn the kinds of returns that First Solar no doubt hoped it would early in its lifespan. In 2017, First Solar wrote down $25 million of its investment in the plant, chalking it up as a business loss. In explaining its decision, First Solar CFO Alexander Bradley stated that the company took the write-down on the project because "declines in retail power prices since the completion of the project in 2014 have resulted in ongoing operational losses that necessitated the write-down in value." Translated into plainer English, this means the plant did not make as much money as First Solar wanted, to such an extent that the company doesn't have hope for improvement and as a result declared a $25 million loss on the project in its tax filings.

While a write-down is certainly an indication of the poor performance of an asset, it is important to be clear about exactly what it does and does not mean. Writing down an investment in a project is not the same as the project going into bankruptcy or being terminated and disassembled. First Solar CEO Mark Widmar explained this distinction himself, categorizing the decision as primarily an accounting issue: "We'll continue to sell the power that's being generated off that asset, but from a book-value standpoint, we had to write it down."[25] In summary, the project was not clearing the financial hurdles that would make it a success but was making enough money to keep operating.

If Barilla has failed to reach its financial goals, does that also mean it is not the marker of solar's newfound competitiveness that I have been arguing it is? Obviously one answer to this question could be "yes": Barilla as a profitable merchant power plant was a concept put forward into the world, and it turned out to be a mistake, plain and simple. Ipso facto, the Solar Age is not upon us. And yet as definitive as that answer seems, it also seems facile in light of the larger situation in which Barilla found itself. There are other ways to look at the question, ways in which the answer is both clear and very different. If you look closely at the years prior to 2014, when the Barilla project was financed and constructed, and then at the years 2015 to 2017, up to the point when the write-down was taken, there are reasons to believe that the project's poor performance may have been more the result of a special set of circumstances than the technology's lack of competitiveness.

In markets such as ERCOT, where natural gas is the predominant form of generation, the price of electricity is determined not only by the supply of and demand for electricity in the market but also by supply and demand setting the price of natural gas. In West Texas, where Barilla generates and sells its output, both of these supply and demand factors converged from 2011 to 2014, when the planning and approval work for Barilla was occurring, to maintain a stable and attractively high power price for electricity in West Texas. Several factors contributed to this situation. Significant new oil and gas drilling activity in West Texas, driven by oil prices that had doubled since 2008, created unusually strong electricity demand in the region over these years. Over the same period, natural gas prices tripled, raising electricity prices across all of ERCOT.[26] Both of these factors exerted upward support on power prices in West Texas, prices which many believed would persist for several years.

Then, as Barilla made its debut in 2014, both forces unexpectedly reversed course, leading to a collapse in the price of electricity. Oil and gas activity declined in the region when oil prices tumbled from $105 a barrel in 2014 to less than $30 a barrel in early 2016.[27] At the same time, natural gas prices fell back to where they began in 2012.[28] Electricity prices in West Texas, which had ranged from $33 to $47 per megawatt-hour between 2010 and 2014, fell to $26 per megawatt-hour in 2015 and then $22 in 2016.[29]

Clearly the investment case for the Barilla plant was put forward for approval in an environment in which market forces painted a picture very different from the one that evolved later. After it was constructed, these many circumstances conspired to weaken the project's performance significantly.

In considering the question of whether Barilla really is a failure as a marker of solar's competitiveness or instead just a victim of market forces, it is useful to step back and consider the fate of another power plant built in ERCOT around the same time. Barilla was not the only power plant to see its prospects seriously deteriorate after 2014; others also suffered as a result of the above factors. One plant—a brand-new, state-of-the-art natural gas power plant built by a highly experienced independent power producer—shared much with Barilla in terms of the market it expected, as well as the one it actually encountered. This project went bankrupt, a fate far worse than Barilla's write-down.

Known as Panda Temple I, this gas-fired plant was named for its location near Temple, Texas, and the company that developed it, Panda Power Ventures. Temple I and Barilla were each planned and constructed between 2011 and 2014; each came online within months of the other in 2014; each received some fanfare for the novelty of its financing; and each took a merchant approach to selling the electricity it generated. Obvious differences were the generation technologies and the size of the plants. At 758 megawatts, Temple I is vastly larger than tiny Barilla, and it generated power from natural gas.

Both plants failed to perform financially. Explaining the reasons for the Temple project's bankruptcy, Panda's Vice President for Public Affairs Bill Pentak was direct in an interview with the local newspaper: "The power prices are very, very low and it's putting stress on a lot of Texas generators. That plant is a great plant, and we would foresee that it would continue to operate. We've got people who are getting debt financing so that it can keep running."[30] Panda worked to refinance the project so it could emerge from bankruptcy and continue operating.

In understanding the dramatic changes in electricity pricing that occurred in ERCOT between 2011 and 2017, and in recognizing that these changes were so significant that they harmed not only Barilla but also fossil generation plants like Temple I, one reaches a different conclusion about Barilla's success or failure. If Barilla's fate is a sign that solar is broadly uncompetitive, should we draw the same conclusion about natural gas power plants from Temple I? Surely not. No one would argue that Temple I's financial troubles marked the day natural gas power plants could not compete on the market. It is much more reasonable to conclude that both plants were likely commercially sound projects for the market they understood to exist in 2014, and that both were overturned by the surprising market conditions that emerged afterwards.

■

Perhaps the best test of whether Barilla truly represents a shift in solar's ability to compete in the open market is to examine whether other merchant renewable energy projects are advancing behind it. If Barilla has been the only merchant renewables project and it has failed, that would suggest the concept is not viable. If it has been the first of many, that would mean something very different.

In the United States, the same unusually low natural gas and electricity prices that challenged Barilla have persisted, slowing solar's growth here to a degree, but the case for merchant solar is advancing nonetheless here and in many other places around the world. Global consulting group Deloitte wrote in June 2020, "After almost a decade of continuous improvements in cost-competitiveness, onshore renewables are now consistently cheaper than conventional power sources and moving into all-merchant territory."[31] As if on cue, Cutlass Solar, a large Texas project at 140 megawatts, was announced in early 2021 as "one of the largest . . . merchant solar" projects ever to be constructed,[32] and several other new projects are said to be moving in the same direction. Big Oil titan BP recently announced that it is banking on merchant solar in its latest return to the renewable energy sector.[33]

To be sure, most solar projects in the US still require power purchase agreements in order to be financed, but these agreements have been changing—specifically, they have been becoming more merchant-like. The number of years that many PPAs last is getting shorter and shorter, increasing the merchant period of the projects' lifespans. Whereas in the past such agreements remained in place for fifteen or twenty years, today they can last as little as seven or even five years. This has resulted in the share of many solar projects' lifetime revenues that occur during the contracts getting smaller and the merchant share occurring afterwards getting larger. Recent estimates are that the newest solar projects are only getting about 15–20 percent of their revenues through the power purchase agreement, and the remaining 80–85 percent comes from merchant revenue.[34] In 2020, the first US solar projects to put hedge agreements in place—a kind of PPA-lite structure that presents much of the same risk as merchant projects—were built, including the 200 megawatt Holstein and 240 megawatt Misae solar projects.[35] Many expect hedged solar projects to increase significantly in the US.

Internationally, merchant solar has advanced in a number of places in the last several years. Chile has been a leader, having seen more than eleven merchant solar projects constructed, including a few well over 100 megawatts in size.[36] Mexico has seen at least three merchant projects constructed recently, and Australia, Canada, Spain, and Denmark have now been added to the group. In Australia, the Yarranlea 121 megawatt, Kidston 50 megawatt, Chinchilla 20 megawatt, and Northam 10 megawatt solar projects were all constructed merchant in 2018, with more having been built each year since.[37] Canada had its first merchant solar project, the 25 megawatt Innisfail, finished in 2019 and is following up in 2021 with the huge 400 MW Travers merchant project. In 2019, the 79 megawatt El Bonal project was announced in Spain and the 155 megawatt Vandel 3 project in Denmark, both merchant solar.[38] Solar developers and bankers speculate that substantially more solar will be built merchant in Europe in coming years.[39]

Solar has not been alone in this regard either; wind is also getting into the game. In Texas and other US markets, some wind projects that have seen their power purchase agreements expire have declined to renew or seek new contracts, opting instead to operate merchant.[40] For the first time ever, in European markets offshore wind energy is projected to beat prevailing wholesale market prices,[41] setting the stage for merchant offshore projects. Onshore wind projects are also going merchant in Europe: Renovalia, the same company that built the El Bonal merchant solar project, announced its plans in 2020 for 1,000 megawatts of new wind and solar projects in Spain, all merchant.[42]

All these occurrences make Barilla look more like the signpost of change than its write-down would suggest. Reasonable minds may differ about the degree to which the merchant model is already taking hold, as well as whether each of the other important qualifications discussed above might diminish the weight of Barilla's claim to being the sea change in the quest for truly market-competitive solar pricing. At the very least, however, there are significant reasons to believe the project was built in a first-ever moment of confidence about solar-generated energy being so cheap that it can stand on its own two feet in a market dominated by natural gas and coal generation. Time will dispel all doubts, one way or the other, and for my part, the collective weight of the many arguments in favor are overwhelming.

Even apart from the question of whether or not more merchant solar plants are being built, I have a simpler way of knowing that solar has emerged as the dominant power-generation technology. Almost all the colleagues with whom I have worked over the past twenty years have stopped developing wind projects and are now working only on solar, and many of the larger companies I've worked with that develop both fossil and renewable power plants are doing more solar work than anything else. This circumstance tracks national trends. According to the U.S. Energy Information Administration's 2019 forecast, if the current pattern continues, over the years 2020–2050 the country will see about the same amount of new solar built as new natural gas generation.[43] If the oil and gas industry sees higher production costs in the future— which could result from proactive policies such as a carbon tax— the U.S. Energy Information Administration predicts more than twice as much solar generation will be built than natural gas generation.[44]

Almost no one, including me, saw merchant solar coming, but it is here now and for the foreseeable future. How did it arrive on the scene and reach the point of being poised for market domination so quickly? In fact, it had been more than 200 years in the making.

PART 2

HOW
SOLAR
HAPPENED

CHAPTER 6

SOLAR ENERGY'S SERENDIPITOUS BEGINNINGS

How a teenager tinkering in the basement discovered photovoltaics, Bell Labs made it work, and Big Oil turned it into a power plant.

The first I knew of solar power was as a boy during summer 1979. I remember my parents pointing out the news that Jimmy Carter had installed solar panels on the roof of the White House. My recollection of this is dim—I was eleven years old—but I remember the panels and that they looked neat, and that it seemed like a good idea to me. I was aware of the Iranian Revolution and the ensuing energy crisis hitting hard that same summer, which caused gas lines to wrap around blocks all over the country. This I remember very well for two reasons. First, my brother and I, desperate to avoid having to cut grass all summer, saw an opportunity as we sat in the backseat in the summer heat waiting for our turn at the gas pump. We decided to fill a cooler with Cokes and go down to the gas lines most afternoons to sell drinks to sweating drivers for fifty cents each. Before long, we both had made enough to buy new bikes.

The second reason I remember those solar panels being installed was that I, just like millions of other Americans, had a dawning sense that our lives rested on energy resources that may not be as secure as we had assumed. Even as an eleven-year-old, I had never seen anything like the gas lines before. The idea was completely new to me that we would not be able to drive the family car to baseball practice because we had not accounted for a wait in line of an hour or two for gasoline. And gas lines were not the only reason people worried about energy in 1979: The Three Mile Island nuclear accident had occurred earlier that spring. This event, which plays an important role as a catalyst in the story of solar in the following chapter, compounded public concern about energy security and the need to develop new power generation technologies like solar.

As it turned out, however, these moments of pondering energy alternatives were fleeting. When the crisis ended, the nation quickly moved on and things got back to normal. The gas lines ended in 1980, and by 1986, the price of oil had fallen by more than half and Ronald Reagan decided to remove the solar panels from the White House roof. Their removal made the national news and sent a clear signal that solar had been just a passing fad and that the "real" energy—oil, gas, and coal—was back.[1]

Solar energy's brief debut in the 1970s is just one step on the much longer path leading to its appearance at Barilla in West Texas. In fact, it was not that long ago that the idea of building solar power plants all over West Texas, or anywhere at all for that matter, was more or less science fiction. As recently as the early 2000s, solar energy was considered an exotic technology, deeply uncompetitive on price, and just plain not ready for prime time— something more suited to a government lab, a hippie commune, or a satellite in space than to doing the gritty work of generating electricity for toasters and coffee machines. The very idea of it was challenging at a fundamental level. How could technology that relied on sunlight, a power source as diffuse as it could be fleeting and intermittent, compete with plentiful, dense, and easily transported coal, oil, or natural gas? Despite sunlight's ubiquitous nature, could it ever match the energy of fossil fuels, themselves hundreds of millions of years of sunlight condensed by photosynthesis and eons of unimaginable heat and pressure into forms already chemically eager to be released by flame and transformed into kinetic energy?

Photovoltaic solar's remarkable evolution from curiosity to science project and then to significant player in power generation has been extraordinary, at times both unlikely and hard-fought. The 150-year story is mostly one of slow, incremental progress, occasionally punctuated by giant leaps forward—leaps that were often the result of startling discoveries or occurrences. Empirical research suggests that between 30 and 50 percent of all scientific discoveries are accidental in some sense,[2] and this has certainly been the case with solar. Chance has intervened numerous times to keep the plot unfolding, often after decades-long waits for new advancements.[3]

■

Photovoltaics—the idea that sunlight, when shining upon a certain chemical, can become electricity—were first discovered in a way that seems apt for the upstart, disruptive technology it has become in the twenty-first century: by a teenager experimenting in his father's laboratory basement.[4] Edmond Becquerel, son of the well-known French scientist Antoine César Becquerel and father to Nobel prize–winning physicist Antoine Henri Becquerel, began his lifelong fascination with all things luminescent when his father enlisted him to assist with his scientific work in the lab in Paris. We do not know the circumstances that led nineteen-year-old Becquerel in 1839 to submerge a certain combination of metal plates in an acidic fluid and then expose one of them to sunlight, but when he did so, he recorded a slight charge emanating from the plates.[5] He established that the light, for a reason he could only guess at, caused the charge to flow. He noted the event in his log, and it is recognized as the first observation of electricity created from sunlight.[6]

While Edmond Becquerel is credited with discovering the photovoltaic effect, he had no understanding of the underlying phenomena producing the charge. Becquerel moved on to other work, and his discovery lay more or less undisturbed for decades. The next step came, like the first, in a completely unexpected and felicitous manner. The English scientist Willoughby Smith had made scientific discoveries during the 1860s that resulted in the first undersea communications cable being constructed across the English Channel. Looking for a way to detect breakages as the lines were being laid, he noticed that the selenium crystals he applied for this purpose functioned very differently when exposed to light than they did in the undersea darkness. He ascertained that sunlight induced some electrical effect in the metal and published a paper with his findings.[7] Smith's discovery was taken up by a number of scientists in the years that followed, including William Adams, a professor at King's College London, and his student, Richard Day. Adams and Day made refinements to Smith's work in their own experiments, which allowed them to conclude, for the first time, that the effect of sunlight on selenium crystals was to produce a flow of electricity. In fact, the sunlight was reacting with the element to create electricity. This was more observation than understanding, but it still represented a major step forward.[8]

This work by Smith, Adams, and Day identifying selenium crystal's unique response to sunlight was then picked up by American inventor Charles Fritts. In 1883, Fritts placed a glass box containing an extremely thin layer of gold over selenium wafers on his sunny rooftop and saw current flow, creating what is now acknowledged as the first solar module. Solar historian John Perlin recounts Fritts' prescient but overly optimistic announcement upon seeing the result: that before too long his "photoelectric plate" would compete with other means of generating electricity.[9]

The new technology was seen as very promising, and a number of scientists began to look into it. Almost fifty articles on solar energy were published in *Scientific American* between 1880 and

1914.[10] Despite this broad level of scientific inquiry, however, the quantity of electricity created from experiments was very small, and understanding about how to advance the science was very slow in coming.

The puzzle moved closer to a solution when Albert Einstein introduced his theory of light in 1905. Many people assume that Einstein won the Nobel Prize (in Physics) in 1921 for his theory of relativity, also published in 1905, but in fact the prize was given for his work on sunlight. Einstein theorized that sunlight, and in fact all light, was made up not of waves but of small packets of energy. His theory was correct, of course, and today we call these photons. This discovery was a huge step forward for photovoltaics, because it provided the basis for understanding how the current observed by Adams and Day might have been created from sunlight.

With the knowledge that sunlight contains Einstein's energy packets, it became evident that some interaction between those particles and the substances experimented with by Day and Adams resulted in the observed electric current. It followed that it might be possible to find a way of converting sunlight into electricity reliably, and that such a conversion would probably depend on the chemistry of the substance interacting with the light—an understanding that has been central to the story of photovoltaics ever since. The advancement of solar energy has been fundamentally an exercise in sleuthing to find the right chemical materials to transform the packets of energy in sunlight into electricity and thus to produce current efficiently. Surprisingly, solar energy then has evolved much more as a chemistry problem than as an electrical engineering one.

Coincidentally, a separate vein of scientific work investigating anomalies in electrical properties of different chemicals had been ongoing since the late 1800s, and this work resulted in the next big step forward for photovoltaics. In 1874, the German scientist Ferdinand Braun observed that, when he touched a charged wire to a lead sulfide crystal, the resulting current flowed in only one direction—not, as expected, in both. Braun had discovered a trait of what would come to be known as semiconductor material.[11] In Braun's time, the effect he discovered was considered a curiosity,[12] but later scientists imagined a vast array of useful applications for materials with such properties. In 1901, semiconducting materials were used for the first time to detect radio waves, and by the 1920s, the amplifying effects of semiconductors had been discovered, which led in turn to the first theorizing about the possibility of transistors.

Work by Braun and others on semiconductive materials was key to the pivotal next phase of work on solar, occurring over the period 1930 to 1960. These years saw transformational advances in photovoltaics, the scale of which is rivaled only by those that occurred from 2008 to 2016 (covered in the next chapter). Nearly all this early work occurred in the legendary Bell Labs facility, located in suburban New Jersey, which has been called "the most innovative scientific organization in the world" in its time.[13]

In 1930, photovoltaic technology was simply an anomaly observed when certain substances were exposed to sunlight. By 1960, after a series of Bell Labs scientists had chanced upon discoveries and built upon each other's work, the technology became something that does not look or work very differently from a modern photovoltaic panel. Bell Labs' prolific stable of collaborating and competing scientists made world-changing scientific discovery after world-changing scientific discovery, resulting in inventions that still reverberate in modern technologies ranging from radio astronomy devices to cell phones.[14] Two of these discoveries were the first transistors and solar cells. Although fundamentally different in application, these discoveries are linked in their reliance on the anomalous electrical properties of silicon and other materials.

One Bell scientist, Russell Ohl, discovered the first silicon solar cell while working in the 1930s on the difficult matter of radar detection. Ohl was a big believer in the promise of silicon and other semiconductors as a way to expand radio applications. He was in college when he heard a radio for the first time: the SOS of a ship being attacked by a German submarine during World War I.[15] In 1940, Ohl noticed very peculiar current flows when light was shone on a cracked silicon slab. He observed that a crack and impurities in the crystal silicon somehow created conditions in which the light shining on the slab resulted in a flow of electrons along a junction at the seam, creating electricity out of sunlight. It was a very small amount of electricity, similar to those that had been observed previously with selenium materials, and the exact role of the impurities in the silicon was not well understood—but historians credit Ohl with having advanced photovoltaics by inventing the first solar cell.[16]

In 1947, other Bell scientists took Ohl's work, in particular his observation of a positive and negative junction in a silicon cell, and invented the lab's most famous device: the transistor. The transistor is widely considered to be the most important innovation of the twentieth century, an assertion underscored by its overwhelming relevance to human existence today. Consider, for instance, the more than one million separate transistors on each square centimeter of the microchip in the computer I am typing this on or any device you are reading it on, each one switching on or off to communicate its bit of information as often as every 0.000000001 seconds.

In this way, research into radio waves at Bell Labs in the 1930s led to the invention of the solar cell in 1940; research on the solar cell in the 1940s then led to the invention of the transistor in 1947; and research on the transistor in the 1950s revealed the next solar innovation.[17] In 1954, Bell colleagues Calvin Fuller and Gerald Pearson were experimenting with transistors made from silicon as an alternative to germanium, which had been a notoriously temperamental medium.[18] Fuller and Pearson noticed that when certain impurities were introduced into pure silicon and conducting metal junctions were then attached in a certain way, electrical pathways were created that facilitated the outflow of electrons dislodged by external energy sources. This discovery was unexpected, particularly since the external energy source causing the flow was light from a nearby lamp.[19] What's more, the resulting current flowed out of the silicon and into the metal contacts fixed to it, resulting in measured current through these wires. Quite by accident, Fuller and Pearson had improved Ohl's first solar cell. However, perhaps because it was neither entirely novel—the solar cell had already been invented by Ohl—nor the objective of their research, Fuller and Pearson just noted the finding and continued their transistor work.[20]

A final serendipitous development, in 1954, led to the next big step in Bell Labs' solar work. Physicist Daryl Chapin had been tasked with developing a power source for Bell's telecommunications hardware located in remote locations. Batteries and small steam engines were feasible for this purpose but had significant limitations, particularly in hot and humid regions, and Bell was working to find better solutions. Chapin was investigating the deployment of solar cells, but this seemed unlikely to work because the output of selenium and silicon cells was so small. However, Pearson and Chapin were acquaintances at Bell, and Pearson shared with Chapin the results of his recent work on the silicon cell interspersed with gallium impurities. Chapin changed his focus from selenium and pure silicon, moving rapidly forward with several refinements to Pearson's findings that had the effect of increasing electric generation dramatically. Chapin's new cells reached almost 6 percent efficiency that same year, a 600 percent improvement over the selenium cells and an even larger increase relative to Ohl's original silicon cells.[21]

From this point forward, Chapin's impure silicon cell approach became the focus of virtually all photovoltaic inquiry. Electricity flow at 6 percent efficiency wasn't much, but it was satisfactory for certain real-world applications. Bell Labs and other technology developers quickly moved, in 1955, to start commercializing "solar batteries" utilizing Chapin's new cell design. One of the first commercial uses came, surprisingly, in another rapidly evolving technological area—the space program. In space, combustion is not possible, so satellites can only be powered by nuclear or solar power. Solar quickly emerged as the more viable option, and in 1958 the United States launched the world's first solar-powered satellite, Vanguard 1.[22] Satellites equipped with solar panels, even the low-efficiency panels then available, meant they could communicate data back for years instead of just days.[23] Thereafter, vast research and development investments in space-program applications accelerated the pace of lab work to improve the photovoltaic technology starting during the 1960s and continuing through the 1990s.[24]

Photovoltaic technology building upon Chapin's design enjoyed a period of steady, if modest, growth starting in the 1950s and continuing through the end of the century. Scientists all over the world took the landmark work completed by Bell Labs and ran with it. Each decade from the 1950s to the early 2000s saw incremental advances across a range of cost and efficiency components—in chemistry, manufacturing practices, and associated technologies (such as the development of power inverters, critical components needed to condition the electricity produced by solar panels for grid use). These decades of improvement increased efficiency and reduced the cost of silicon cells and integrated projects. Improvements were also seen in alternative chemistries for generating electricity from sunlight, such as thin film, which applies new, non-silicon photoreactive compounds directly to glass to create solar cells.

All this progress led to fantastic reductions in production costs over this period that were nothing less than mind-boggling. All in all, costs fell from $286 per watt in 1955 to $100 per watt in 1971, then down to $11 per watt in 1980, $7 per watt in 1985, and finally $5.30 per watt by the 1990s. In total, the cost of a watt of solar energy fell a staggering 5,000 percent between 1955 and 1995.[25]

■

One result of Bell Labs' transformational work on photovoltaic technologies between 1930 and 1960 was that solar cells began to show promise for broader applications as an electricity generation resource. In another instance of remarkable serendipity benefiting the story of solar, global events of the mid-1970s conspired to catalyze exciting new deployments for the technology, leading to the progress of photovoltaics outside of the laboratory finally catching up to the progress inside. Over the next thirty years, solar panel usage became commonplace, first on satellites and at remote locations and industrial installations—including the White House roof—and then, in the 1980s, in the first-ever utility-scale photovoltaic power plants.

For most older Americans today, the 1970s energy crises are likely but a distant memory, but when they were occurring it was a difficult and traumatic time. The first shock occurred in 1973, when OPEC (Organization of the Petroleum Exporting Countries) countries placed an embargo on the United States and European nations for supporting Israel following the Yom Kippur War, and the second occurred after the Iranian Revolution in 1979. Both events caused the price of gasoline to increase many times over in a very short time and, even worse, spurred outright shortages of fuel at many gas stations. Oil sold for about $3.50 per barrel in 1972 and had nearly tripled to $9.50 per barrel in 1974 after the OPEC embargo. In 1978, the price reached $15 per barrel and climbed again to $37 in 1980 after the Iranian Revolution.[26]

The result was the scene described at the start of this chapter: cars parked around the block for hours at a time, waiting for a chance to fill up their gas tanks, and a great deal of unprecedented anxiety about the nation's energy supply. These events revealed not only the degree to which Americans had taken imported oil for granted but also the incredibly paltry range of energy options available in the market to meet the nation's needs. The country's energy policy turned, in particular, to domestic resources immune to manipulation by hostile trading partners.

These energy crises were contemporaneous with another 1970s phenomenon, the rise of the environmental movement. What began with Rachel Carson's landmark book *Silent Spring* in 1962 had matured several years later into a robust social movement with a strong public policy component to it.[27] In 1970, the Environmental Protection Agency was formed and the Clean Air Act passed, and two years later landmark amendments to existing legislation created the Clean Water Act. Exposure of energy supply vulnerabilities arising from the oil embargoes provided the movement with another compelling reason, beyond air and water pollution, for Americans to rally around clean-energy sources such as solar and wind. The result was a raft of legislation supporting advances in solar, from requiring solar panel deployment in public buildings to providing substantial new funding for public research. One of the main thrusts of the new support was commercialization of solar, including formation of a new agency, the Energy Research and Development Administration, with this specific focus. In 1978, new tax credits were passed to encourage the use of commercial and residential solar.[28]

One of the most surprising, and consequential, outcomes of these two interrelated trends was the entry into the solar business of several major players in the oil industry. While their motives were sometimes doubted—conspiracy theorists wondered if the oil companies' secret plan was to buy solar technologies just to bury them in order to vanquish a competitive threat, something they may be guilty of regarding other technologies—it seems much more likely that these ventures were earnestly undertaken on commercial bases to diversify, should the oil crisis continue indefinitely. Undoubtedly, the oil companies also valued the substantial new state and federal incentives applicable to alternative energies and research and development spending.

Between 1973 and 1979, Exxon, Mobil, ARCO, and Amoco each acquired major solar technology companies and advanced them significantly.[29] These companies and others that became involved brought not only substantial research and development dollars to solar energy but also a savvy business approach to deployment of the technologies in the field. These ventures were focused on commercializing the technology in the energy space, which represented something of a change at the time. So common was the use of solar panels in space-program applications in this era that the new direction pursued by oil companies intent on commercializing the technologies closer to home earned its own name: "terrestrial photovoltaics."[30]

The major oil companies' ventures into solar were productive to varying degrees, with one of the most notable being that of Atlantic Richfield Company (ARCO). ARCO had been distinguished for its relatively friendly environmental approach to the energy business, making it a natural partner in the solar sphere.[31] In 1977, ARCO acquired a struggling but promising solar company for about $300,000 and invested over $200 million in it over the subsequent ten years.[32] ARCO Solar, as the venture was called, broke the mold for solar ventures in many ways. In terms of manufacturing capabilities, it racked up many firsts in the industry, eventually accounting for 15 percent of the entire solar panel market worldwide. Led by its visionary founder, Bill Yerkes, the company laid the foundation, according to one industry veteran, for today's cost-effective photovoltaics manufacturing.[33]

ARCO Solar was also responsible for breaking the science experiment / remote power source barrier that had constrained the industry up to then. In 1982, ARCO Solar built the world's first grid-interconnected, utility-scale solar project: the one-megawatt Lugo Project, located near Hesperia, California. Unlike every prior solar project, executed to fuel some specific power need located adjacent to its panels, Lugo functioned like any other power plant—by generating electricity and delivering it into the grid in the very same way as any coal, nuclear, or natural gas power plant. Lugo's electrons were indistinguishable from any others, powering washing machines, air conditioners, and everything else in Southern California.

Lugo was also the world's largest solar project at the time, its single megawatt of generating capacity making it three times bigger than any other then existing.[34] The project's 108 arrays were mounted on axis trackers, allowing the panels to follow the sun all through the day. Altogether, the project generated enough power to cover the needs of 300 to 400 homes.[35] The following year, ARCO did itself one better, building the Carrizo Plains solar project at 5.2 megawatts in size, which held the record for largest project for many years. Output from both plants was sold to California utilities under long-term agreements.

Then, suddenly, the advance of the oil companies into solar stopped as abruptly as it had begun. Oil markets finally stabilized, then prices fell dramatically in the early 1980s. When the price of oil plummeted from $30 per barrel in 1983 to $15 in 1986, concerns about the long-term oil supply vanished and oil executives returned full time to the business of drilling and extraction, heading for the exits in their solar ventures.[36] ARCO sold its solar business to Siemens, and the Lugo and Carrizo projects were sold to private investors. A distant offspring of the ARCO Solar business continues to this day as Solar World—a company that plays another dramatic role in the continuing solar story—but by the mid-1980s, the engagement of oil companies in the photovoltaics business had evaporated.

Notwithstanding this fact, ARCO's brief foray had been remarkably consequential, even transformative. The oil-backed ventures are credited with pioneering technologies presaging the massive price reductions achieved in the era,[37] but this wasn't their biggest impact. ARCO's involvement in particular altered the business model of the solar industry in fundamental ways. Photovoltaics were not just research or space projects anymore; they had proven themselves capable of playing in a whole new arena: generating and delivering electricity into power markets alongside traditional power plants. Where space had been solar's primary market in 1975, by 1983 it also had "terrestrial" markets in its sights, an opportunity both vastly larger and more accessible. While the oil companies' stay had been fleeting, it can be argued that it was long enough for Exxon, Mobil, ARCO, and the rest to stamp their competitive business DNA on the industry, which had theretofore been predominantly scientific and academic. This would prove to be a profound and deeply consequential step in the evolution of the industry.

■

Solar progress since 1950 had been breathtaking, but notwithstanding the sea changes that had occurred in lowering costs and improving efficiencies in the turn of the century, solar was still not ready for prime time as a new player in the electricity business—not by a long shot. As a potential source of competitively priced wholesale market power, solar remained a niche technology. As much as production costs had fallen, by the year 2000 solar-generated electricity was still stratospherically expensive compared to the market prices of the day. I had just begun my career in renewable energy a few years before, and while wind was emerging as a competitively priced option, solar was on no one's radar. In wholesale power market terms, the cheapest solar energy available required a price of about $250 per megawatt-hour to turn a profit, but prevailing electricity market prices were five times lower, around $50 per megawatt-hour. Yes, the price of solar was much cheaper than it had ever been and, yes, in certain applications solar cells were the most price-competitive option available, but if you were focused on achieving a widespread deployment of panels into the markets, there was little reason to even consider it as an option. In 1999, installation of solar panels *worldwide* amounted to only 1,000 megawatts total—roughly the same amount of power generated by just two coal plants. Virtually no US utilities were even considering introducing solar into their generation mixes, and it was hard to blame them, given how expensive it was.

All of this was about to change in a shockingly short period of time. A little more than ten years later, all the utilities to whom I had been trying to sell wind power would begin looking hard at solar power options. Another few years after that, solar energy was all many of them wanted to buy.

CHAPTER 7

THE SOLAR JUGGERNAUT

UNEXPECTED OUTCOME OF AN UNPLANNED ALLIANCE

How three disparate events—Germany's improbable pivot to solar, the Great Recession, and China's unprecedented response to it—reinvented solar and birthed a juggernaut.

For full disclosure: I spent most of 2000 to 2015, the next period of solar's history—its most remarkable phase, the one in which it was transformed into the worldwide juggernaut it is today—believing it would never happen. I remember attending conferences and hearing solar companies proclaim their prospects and honestly feeling badly for them. It seemed hopeless to me, given all the sky-high solar prices I was hearing about.

The moment of truth finally came for me, and I learned the hard way that solar had arrived. A wind project we had been developing for several years in Connecticut was beaten in a bid in 2013, not by another wind project but by a solar one. This was the first of several such experiences we had as solar's advance gained momentum. In 2014, the year Barilla was built, things had changed so much that my company decided to convert a wind project we were developing in Maryland into a solar project because doing so would allow us to bid a more competitive price. By then, the market had moved so much in solar's favor that once we switched the project to solar, we quickly found a buyer.

All of us who were slow to acknowledge solar's emergence as a powerful and competitive new power generation player, capped by the appearance of Barilla in 2014, can be forgiven for not seeing the signs, so quickly did they arrive on the scene. Solar's sudden ascent is the direct result of the astounding transformation of photovoltaic technology that occurred between the years 2000 and 2014, which, in turn, could not have happened without the multiple innovations from the previous 150 years recounted in the last chapter. The twenty-first-century transformation of solar has seen it evolve from the kind of technology that only made sense to use when no other power source was available to what it is today: the fastest-growing power generation technology worldwide, and one that has been challenging, and beating, not just wind but also fossil fuel plants in a growing number of markets around the country and world.

During the years 2000 to 2014, a dynamic began that fundamentally reshuffled solar's place in the energy world. The wholesale power market price of solar energy fell precipitously, from about $250 per megawatt-hour in 2000 to $100 in 2010, and then to $50 in 2013. Recently, this price has fallen even further, to below $25 in some markets in 2019, and there is every reason to believe it is headed even lower in future years.[1] Not only have panel prices continued to fall over these years, panel efficiency has simultaneously increased sharply, rocketing from single digits in the 1990s to the high twenty percent range by 2015 to nearly fifty percent by 2020,[2] and improvements have also been made in project financing options.

As prices have fallen, photovoltaics have been deployed in a dizzying array of new places, from residential rooftops to commercial and factory facilities and, most importantly, they are also being used in more and more utility applications of great scale. To get a sense of how jaw-dropping this growth has been, consider that between 1999 and 2013, total solar panel installations worldwide increased a hundredfold, from 1,000 megawatts to 100,000 megawatts (or 100 gigawatts). From another vantage point, consider that the 1,000 megawatts that existed worldwide in 1999 was the same amount of solar generation being installed *each month* in the United States since 2015.[3] Cumulative installation of solar worldwide, meanwhile, more than doubled between just 2013 and 2015—increasing to nearly 250 gigawatts total[4]—and nearly tripled that figure in the four years since, reaching 627 gigawatts in 2019.[5] As a point of reference, Texas uses about 80 gigawatts of power, meaning that total solar installed worldwide in 2015 amounted to about three Texases' worth of generation, and by 2020 reached about eight Texases' worth across the globe. Forecasts project no halt in momentum: global installations are projected to hit as much as 150 gigawatts of new solar per year by 2023.[6]

Something extremely significant occurred between 2000 and 2015 to cause all this, something that took the vast scientific advancement of photovoltaics during the previous century and propelled the technology into unprecedented commercial viability. The events that led to this transformation were a series of interlinked actions and reactions among three countries: Germany, China, and the United States. While the interlinked events creating the situation were complex, the broad outlines of the result are not: a near gifting of the most advanced Western photovoltaic technology, expertise, and personnel to China; gargantuan, unprecedented levels of Chinese governmental subsidization being committed to solar production; and a tidal wave of cheap solar panels being exported from China to the United States and Europe, decimating competitors and dropping the price of solar energy through the floor.

The circumstances that led to this outcome were implausible. Who would have guessed that Germany, the United States, and

China would ever collaborate in areas as contentious as technology, jobs, and subsidies, and do so to make a renewable energy source market dominant? In fact, it can be argued that there was barely any collaboration, merely a pursuit of self-interest by each country that happened to result in this outcome. Whatever one wishes to call it, the effort undoubtedly involved the sharing of the prodigious and unique strengths of each party, which has accomplished a mutually acknowledged important result. If the resulting cheap solar really does introduce the Solar Age and bring about a chance to significantly reduce the harms of climate change, history will probably look back upon this unlikely and unplanned alliance that arose out of these interrelated events as the turning point.

■

The events causing solar's transformation in the 2000s began almost twenty years earlier with a catastrophe having nothing to do with solar energy. In April 1986, just outside the town of Pripyat, Ukraine, workers in a nuclear plant were attempting to test an emergency cooling capability for the plant when things went terribly wrong. Because of a combination of human error and a plant design flaw, a steam explosion ripped the cover off the reactor and the materials within burned for ten days in the open air. Of the 190 metric tons of fuel and fissile materials inside, somewhere between 13 and 30 percent escaped into the atmosphere, contaminating a wide area.[7] This tragedy at the Soviet Union's Chernobyl plant, coming not even ten years after the Three Mile Island accident in Pennsylvania, was a strong sign to many around the world that nuclear power was not going to be the plentiful, problem-free energy resource it had once promised to be.

Although the majority of the radioactive materials spewing from Chernobyl fell on the Soviet republics of Ukraine and Belarus, as weather conditions shifted some of it also spread across Western Europe, including over Germany. In Germany, there was special sensitivity to this disaster because the country had a long tradition of strong opposition to nuclear power.[8] An early example of many Germans' negative views on nuclear power was the successful effort, in 1973, to oppose a nuclear plant proposed for construction in Wyhl, West Germany.[9] Conservative farmers in the area worried about harm to crops and human health, and they joined with left-leaning urban and university protesters to occupy the building site. The project was canceled. Other battles opposing new nuclear plants were lost, but the victory at Wyhl galvanized opponents by proving that strong, consistent, and informed grassroots opposition could block the nuclear plans of even the largest German utility. The movement grew over time, and after Chernobyl, not a single new nuclear power plant was built in Germany.[10]

The Chernobyl disaster energized Germany's strong pro-environment, anti-nuclear protest movement, which had also achieved a big political success just a few years earlier when Germany's new Green Party won seats in the national legislature.[11] Opposition to nuclear power was a major plank in the Green Party's platform, which grew out of postwar pacifism, the antiwar movement of the 1960s, and the tense reality of the Cold War that came from Soviet and NATO (North Atlantic Treaty Organization) nuclear missiles staring across at each other from East and West Germany from the 1950s through the 1980s.[12]

The year 1986 saw another landmark event that further energized the Green movement in Germany. That year, *Der Spiegel*, a popular German newsmagazine, ran a cover story on climate change featuring an image of Cologne's famous cathedral partly flooded by rising oceans. German reaction to the image and the cover story was strong around the country and focused attention on greenhouse gases coming from power plants long before many were concerned about the issue.[13]

The result of the Chernobyl disaster, the *Der Spiegel* article, and the Green Party's first election victories was a growing base of political support for both an anti-nuclear and a Green agenda.[14]

This was reflected in some significant legislative victories, like an innovative 1990 solar roof program and a 1991 law requiring that all utilities purchase a portion of their generation from renewable sources at fixed prices. Then, in the remarkable elections of 1998, the Greens won so many seats that they became a significant player in German politics and were able to form a governing coalition with the Social Democrats. One of the first things the new government did was pass a law requiring the complete phasing out of nuclear power from Germany; following close behind was a major initiative to promote wind and solar energy in creative and meaningful new ways. In this matter, the chain of events leading to the Green/Social Democrats governing coalition resulted in two powerful drivers that paved the way for explosive growth in renewables in Germany: reduction of existing energy supply arising from the nuclear phaseout and major policy support for new wind and solar. The fact that all this occurred even though Germany isn't very windy or sunny is a sign of how strong the governing coalition's commitment was to renewables.

The new government launched several programs that rapidly grew solar in the country. An earlier plan, called the 1,000 Roof Program of 1990, had been a success, so in 1999 the 100,000 Roof Program was introduced, with government subsidies available to homeowners to promote its implementation. Utilities were further required to purchase renewable energy—both wind and solar—at rates fixed by legislation calculated to be high enough to stimulate growth. These programs were wildly successful in terms of increasing solar installations. By 2008 it was estimated that Germany had 500,000 solar roofs in place, and the utility program had grown so much that its cap was removed to encourage still more growth.[15] As early as 2003, Germany's solar roof programs had made it the main driver of international solar panel demand.[16]

The result was, not surprisingly, such a huge growth in demand for panels that Germany's domestic manufacturers had trouble keeping up with it. These German solar panel manufacturers were some of the best in the world at that time but faced various constraints, particularly due to shortages in silicon supply, which drove up their costs and made it difficult for them to grow fast enough to meet demand.[17] As a result, foreign manufacturers—American and Japanese, mainly—were coming to fill more than half of the overall German demand.[18] Although Germany had created the world's leading solar market as a result of its progressive energy policies, its own solar manufacturing companies were unable to reap the majority of the benefits themselves.

This set the stage for perhaps the most consequential part of the story of the unplanned alliance, in which the best German photovoltaic companies, with their first-tier expertise, personnel, equipment, and research and development efforts, turned to nascent Chinese solar manufacturers for help. Over a few years' time, through this relationship, German companies ended up transferring a significant part of these assets to their Chinese partners, thereby paving the way for China's ascendance in photovoltaic manufacturing.

By any measure, up to the early 2000s China appeared to be an exceedingly unlikely future world champion of solar panel manufacturing. To the extent that Chinese policymakers had favored any renewable energy technologies at all in their industrial and infrastructure planning up to this time, they had focused on wind and hydroelectric energy. Prior to 2006, solar was simply not a priority among Chinese private companies nor among central government planners, who generally show no compunction in picking winners and losers. Lists of the top solar manufacturers worldwide confirm this: not a single Chinese company was included for the years 1988 and 2001, and just one company for the year 2006.

The Chinese solar companies that were active in the early 2000s faced a daunting landscape. The internal Chinese market for solar at that time was small and not expected to grow much, so most of the opportunity lay abroad, in markets dominated by large and highly advanced foreign companies. These companies were fierce competitors with complex manufacturing capabilities, prodigious intellectual property, experienced management, and good access to capital. The idea of a group of small, recently started Chinese companies taking over this industry within just a few years would have seemed an absurd proposition at the time.

To begin to establish themselves, freshly minted Chinese companies such as Trina, Yingli, Suntech, and the creatively named Canadian Solar adopted a strategy to hire away foreign-trained Chinese executives from North America and Australia and to import manufacturing machinery—the machines that make the solar panels and inputs—from Europe and the United States.[19] Even with hired guns in management and the right equipment, however, they still faced the soft but very real business disadvantage of being "new," and all that comes with it—including questions from potential customers about quality control and execution capability, the kinds of questions that would not be asked of the already proven German, Japanese, and US solar companies.[20]

When German companies came knocking on China's door in the early 2000s looking for solutions to their problems—fundamentally, a cheaper supply of solar panels they already knew well how to make—the young Chinese companies that eagerly answered were likely a good enough fit, with plenty of access to cheap inputs but limited know-how. The German companies decided to take a chance on them, entering into agreements to provide equipment and expertise in return for manufacturing relationships for supply of needed components. The Chinese companies responded by rapidly accommodating the orders, raising capital quickly to grow their operations as business opportunities grew, and facilitating development of a supply chain within China to support the new operations.[21]

Through these initial relationships, German companies utilized the expanding capabilities of their new Chinese partners to meet rising panel demand. In doing so, they helped the Chinese companies enormously, consolidating their positions as credible players on the world market. Germany became the largest supplier of highly sophisticated manufacturing equipment to China over this period, providing, along with the equipment, consulting services that supported the development of the expertise needed to operate it. German companies also increasingly entered into joint ventures with Chinese companies, which facilitated the even deeper transfer of expertise and helped their Chinese partner companies learn how to obtain things like the valuable third-party quality certifications that customers expected.[22] The Chinese companies also benefited from goodwill as a result of their growing interactions with their German partners. If their manufacturing was good enough for blue-chip German solar companies, then surely it was good enough for other leading solar companies as well.[23]

The expanding relationship, overall, saw both sides benefit greatly. For their part, German companies profited by reselling cheaper and more plentiful panels to their customers in Germany, while the upstart Chinese companies not only gained access to the fastest-growing market in the world—a market they would otherwise have had no hope of entering—they also exponentially enhanced the highly valuable, intangible components of their businesses, their knowledge and expertise and credibility with customers. Germany may or may not have realized the extent to which they were creating a new competitive force in the world photovoltaic market, and they may not have cared much if they did know, so well did the arrangement seem to meet their needs. But the assistance they provided to the upstart companies was likely beyond anything the Chinese companies might have dreamed of.

It wasn't just German companies that helped grow the emerging Chinese photovoltaic manufacturing industry so quickly. US and other external investors, liking what they saw in the new Chinese solar industry's growth prospects and increasing competitiveness, decided to get involved in a big way by investing vast sums in them between 2005 and 2008. An intense round of venture capital investment and IPO activity in this period infused the

Chinese companies with additional cash that helped them expand their activities. In 2005, Suntech raised $400 million from listing on the New York Stock Exchange; in 2006, Trina raised $98 million and Canadian Solar raised $115 million; and in 2007, Yingli raised $391 million.[24]

The growth of the Chinese solar industry from the late 1990s to 2008 was remarkable. It increased from a handful of start-ups to a stable of quickly growing, prosperous companies with international relationships and track records with some of the world's best customers. By 2004, China was well established in several parts of the solar value chain, including the manufacturing of silicon wafers that go into the fully integrated panels themselves.[25] By 2007, the industry had grown its operations to such a size that it was reaping substantial economies-of-scale benefits and seeing its growth path endorsed by the world's most sophisticated investors.[26] In 2003, Chinese panel production was only 35 megawatts; by 2008, Chinese firms were shipping 2,600 megawatts, about a third of all worldwide shipments, making China as a country the largest photovoltaic manufacturer in the world.[27]

The close, formative relationship between the German companies, which were the handmaidens in this part of China's development as a solar powerhouse, and the industry in China has been well documented in academic writings but for some reason has not been widely reported. The author of the most informative works on the subject, Dr. Rainer Quitzow of the Institute for Advanced Sustainability Studies in Potsdam, Germany, has written about the uniqueness of the ingredients of China's ascendance and the serendipity of its path. "Other regions attempting to create a competitive photovoltaic manufacturing base likely could not replicate all the favorable conditions experienced in China during its critical growth phase," he wrote.[28] The key element in this unusual mix of circumstances was that, Quitzow concluded, "Germany's rapid demand growth, set against shortages of polysilicon and thus PV modules, set into motion a unique period of dynamic international interactions and interdependencies, in particular between Germany and China."

China ended the year 2008 with the strongest solar industry in the world, having overcome the challenges represented by fierce established competitors in most of the world's best solar markets: Germany, Japan, and the United States. However, the arrival of the global financial crisis would present another set of challenges, and China's surprising reaction to them would consolidate its position as the world's leading solar manufacturer for the foreseeable future.

■

Anyone who lived through the 2008 global financial crisis, or who watched the movie *The Big Short*, is familiar with the basics of what happened: a cascading series swallowing up bankruptcies and sell-offs sparking an ever-wider sinkhole of shaky assets, triggering a protracted, worldwide recession. That, at least, is what it looked like here in the United States. The picture was different in a few important ways in China. While the American economy deteriorated rapidly from mid-2007 on, the gloom took a while to catch up in China, which was enjoying strong growth through the beginning of 2008. As customers around the world reduced or even ceased purchases of goods across the board, however, the recession began to creep into China in the form of a precipitous drop in demand for its many exports.[29]

For the solar industry, the drop in demand for panels resulting from the crisis was particularly swift and sharp. Spain, which had followed Germany's lead on solar-friendly policies and by 2008 had displaced Germany as the largest photovoltaic market in the world, introduced a strict cap on new solar capacity in 2009 that immediately sent shock waves through the supply chain.[30] Germany's highly successful fixed pricing for renewables, called a "feed-in tariff," was also reduced in 2009 and then again, very significantly, in 2010.[31] France's program was cut sharply as well.[32]

Demand also fell across many other countries in Europe, from Italy and Greece to the United Kingdom and Ireland—all signs of how hard the continent was hit by the financial crisis.

Chinese leaders reacted to the crisis in ways similar to those of other world leaders: moving to stanch job losses, shoring up important institutions, and, most importantly, stimulating growth anew. Renewable energy, including solar, was quickly included in China's robust post-crisis stimulus: in November 2008, a $586 billion package was passed that earmarked a "major portion" of spending for renewables, including solar.[33]

However, it was not only the financial crisis that was darkening the horizon for solar manufacturers at that time. While many energy market observers had read the skyrocketing natural gas prices in 2008 as a trend likely to persist, they were shown to have been terribly wrong when prices collapsed the next year. Given that much of the world's electricity is generated by natural gas power plants, these falling prices led to falling wholesale electricity prices, which meant solar was suddenly also less and less competitive in wholesale power markets. Whereas prior to 2008, purchases of solar energy were often competitive with wholesale power price forecasts associated with ever-rising natural gas prices, this justification became much harder when gas suddenly became very cheap. Reading these many bad signs, solar panel manufacturers in every country began fearing for their economic lives.

China's situation was in some ways particularly precarious. It is worth recalling that, on the eve of the crisis in 2008, China was the largest photovoltaic manufacturer in the world—but also that it exported 98 percent of its production to customers abroad. It was not hard for Chinese policymakers to read the same writing on the wall as CEOs in the European solar companies that were about to go bankrupt were reading: a reckoning was coming for China's juggernaut solar industry. China's near-total dependence on the quickly shrinking European and North American solar markets would soon lead to severe financial difficulties across the sector.

Would China let the market enforce its harsh punishment with a round of Chinese bankruptcies like those that rolled through German, French, and American companies, or would it take some action to prevent this?

There is a school of thought that China, owing to its relatively closed political system and the unique challenges of governing its large and diverse population, puts a special emphasis on the need to maintain economic growth not just for its own sake but also for preservation of political order. In political science circles, the term "performance legitimacy" has been coined for the idea that the stability of China's government relies upon its ability to deliver results to its constituencies, particularly economic growth.[34] While any government ultimately relies upon the consent of the governed, the idea of "performance legitimacy" is that, if you are an unelected government attempting to retain power over 1.3 billion people, that consent is particularly conditional. In a democracy, a government that fails to manage response to an economic crisis loses the next election; under an autocratic government, discontented citizens only have recourse through more dramatic means. A Brookings Institution report from 2009 makes the point as follows: "The existence of millions of disgruntled unemployed workers is a concern for any government, yet there are distinctive institutional features in China that make the regime particularly vulnerable to this threat."

While the rationale provided for any official Chinese policy can be quite opaque, from the "performance legitimacy" perspective many of China's decisions in the wake of the financial crisis would have been driven by a desire for as fulsome a response to the crisis as possible to prevent the very real threat of large increases in unemployment. To the extent that photovoltaics manufacturing presented a way to keep and in fact grow substantial numbers of factory jobs in a sector with a record of strong exports, it would likely have compared favorably with other options under consideration.

Whatever the reason, in 2009, China made two major decisions about solar that altered the course of the industry's development fundamentally. First, it formally identified solar manufacturing as one of seven new "strategic industries" and charged the China Development Bank with implementing support for the industry.[35] China had provided subsidies to solar manufacturers prior to 2009, but around this time support increased substantially. The bank promptly provided about $40 billion in credit extensions to the industry and an additional $30 billion line of credit specifically for manufacturers.[36] China's central government also asked local governments to take steps to support the newly designated strategic industries.[37] This approach had been used to support industries of strategic importance in the past, such as certain heavy and chemical industries.[38] This time the method enabled a new subsidization regime that took a soup-to-nuts approach for solar manufacturers, providing support through the entire business plan: free or reduced-rate land, free or low-cost loans and loans with reduced credit requirements, tax rebates, research grants, energy subsidies, technical and personnel support, and even cash grants.[39] In other words, the full weight of Chinese state resources was now being thrown behind those firms engaged in solar panel manufacturing.

The second step taken was to encourage major construction of new solar generation projects within China for the first time as an additional way to support the industry going forward.[40] Programs were introduced to drive domestic demand for new panels, and other steps were taken to increase deployment within China. These programs were effective, growing domestic solar rapidly from about 300 megawatts in 2009 to a staggering 33,000 megawatts just two years later in 2011.[41] Due to local content requirements, there were few opportunities for non-Chinese vendors to participate in these projects.

The breadth of subsidies available to major manufacturers was captured in part by one researcher's account of Suntech Power Co. Ltd.'s experience. Suntech, at one point the largest photovoltaic manufacturer in all of China, received local bank loans that totaled $56 million in 2005 that grew to $3.7 billion in 2012. The lending bank was state-owned and provided the loans at below-market interest rates upon direction from the local government. Between 2006 and 2011, Suntech also received tax rebates and other refunds totaling $1.42 billion.[42] Incredibly, in the midst of the financial crisis, Suntech more than doubled its annual production capacity between 2009 and 2011, to 2,400 megawatts.[43] Although most major Chinese manufacturers survived the crisis, Suntech was one that was allowed to go bankrupt in 2013, whereupon its market share was absorbed by other Chinese companies.[44]

By 2012, the top ten solar manufacturers in China had accumulated a combined debt of $17.5 billion.[45] The natural consequence of these vast loans, subsidies, and incentives being made available to the manufacturers was, unsurprisingly, a large increase in their production capability. Between 2009 and 2011, production in China increased more than eight times, and China's share of worldwide PV cell output grew from about 30 percent to about 60 percent.[46] In 2012, a year in which just about 31 gigawatts of solar generation were installed worldwide, the combined production capacity of all Chinese solar manufacturers was about 40 gigawatts.[47] The large oversupply also caused solar panel prices to decrease. In 2013, when the financial crisis had largely passed, China had dropped worldwide panel prices by about 80 percent from 2008 prices.[48]

The collective conditions in the international solar panel market in the early 2010s were punishing, not just because of the loss in demand for solar panels in key European markets but also because of the plummeting prices caused by Chinese overproduction. The effect was disastrous for manufacturers everywhere else. In 2011, French solar giant Photowatt went bankrupt, as did US companies Evergreen and Solyndra.[49] The next year saw an epic wave of bankruptcies spread through the German market, hitting virtually every German manufacturer of any size—Q-Cells, Solon,

Solar Millennium, Solarhybrid, Scheuten Solar, and Odersun—all within a few months of each other.[50]

By the time the financial crisis had passed, the unplanned alliance had completely redrawn the solar manufacturing space. The same worldwide ranking of solar manufacturers that had listed no Chinese companies in 2003, and only one in 2006, was by 2011 listing eight Chinese companies in the top ten. This feat was all the more remarkable considering how much the size of the global market had skyrocketed between 2006 and 2011.[51]

In all, China quadrupled its solar panel manufacturing from 2009 to 2011, surpassing virtually all leading companies in the United States, Germany, and Japan.[52] By 2013, China accounted for 67 percent of panel production worldwide and had arrived at a position of worldwide domination of the industry.[53] One recent report shows an even more complete picture of the current Chinese domination of the sector, one that industry analysts do not expect to see change anytime soon.[54] The 2017 report, from Stanford University, surveys China's position relative to that of the United States, a leading solar manufacturer as recently as 2006:

> China accounts for 52% of polysilicon manufacturing capacity, 81% of silicon-solar-wafer manufacturing capacity, 59% of silicon-solar-cell manufacturing capacity, and 70% of crystalline-solar-module manufacturing capacity in the world. The United States, by contrast, accounts for 11% of the world's polysilicon production capacity, 0.1% of wafer manufacturing capacity, 1% of cell manufacturing capacity, and 1% of module manufacturing capacity . . .[55]

Within the space of much less than a decade, US solar manufacturers moved from positions of leadership to struggling for survival, victims of China's multifaceted and overwhelming competitive onslaught. Erstwhile leading companies in Germany, Japan, and elsewhere experienced similar fates.

To achieve this end, China flouted international trade rules—there is no question about that—providing fodder for the Trump administration's "America First" trade stance that prioritized US manufacturing jobs over all else. However, in doing so China paid a price of its own on its way to creating a clean and cheap power generation source that every day becomes more effective at retiring fossil fuel generation—an outcome that can potently advance climate goals far beyond anything else on the scene today. In many ways, the United States is the greatest beneficiary of the unplanned alliance, having secured the ability to purchase a near boundless supply of ever cheaper solar panels that have underwritten the appearance of Barilla and its progeny. Clearly, weighing the damage caused by China's anti-competitive tactics against the climate gains these tactics secured, and the other ways China, the United States, and other countries gained and lost in this story, requires a broader perspective than whether any single country gained or lost jobs.

CHAPTER 8

WINNERS AND LOSERS FROM CHINA'S TRANSFORMATION OF SOLAR

How climate change is scrambling our traditional views of winners and losers when it comes to tariffs and trade.

T he primary narrative regarding China's phenomenal rise as a solar powerhouse has been about its vast subsidies distorting markets, destroying competitors with unfair practices, and costing the United States, Germany, Japan, and other countries important manufacturing jobs. In this scenario, China is the "winner" and the other countries who lost jobs are the "losers." This is not an inaccurate viewpoint, as reflected by the numerous actions taken by the World Trade Organization and others after concluding that improper subsidization has occurred, and it was fair game for the Trump administration to call foul on these practices, just as the Obama administration did before it and the Biden administration is doing after it. From an economic theory point of view, predatory trade practices by nations distort markets and prices and further initiate a race to the bottom, with subsidization wasting public funds and causing real harm to the economic interests of trading partners.

From a climate change point of view, however, a very different narrative emerges. It is helpful to put aside these narratives about trade rules and which country gains or loses manufacturing jobs and consider the basic facts of the entire situation in a different light. A wider view of the exchanges between the various parties reveals a much more complex and nuanced picture, with a broader set of gains and losses for each party. In this picture, it is not as clear who won and who lost, as there are costs and benefits to go around. Consider the following five observations about these pivotal events.

First, in transforming solar so quickly, China did something no other nation, including the United States and other Western countries, were prepared to do. Climate advocates would acknowledge that, whatever it is appropriate to say in defense of fair trade, it is highly doubtful that Western companies could ever have advanced the cause of price-competitive solar nearly as quickly and effectively as the Chinese government has with its unprecedented direct intervention in the solar business. At the end of the day, the massive power of the Chinese government getting behind an industry in a way that Western governments rarely if ever do, and the 2009 Chinese leaders' decision to go "all in" on solar—not only with government stimulus but also with the prodigious if shadowy additional resources of local governments and various public and private banks—were defining factors in transforming solar from science project to worldwide energy juggernaut. "If there was ever a situation where the Chinese have put their whole governmental system behind manufacturing, it's got to be solar modules," wrote Ken Zweibel, a solar industry expert with the U.S. Department of Energy in 2016.[1]

China was uniquely capable of doing so much, so quickly, to reshape a major world industry in a few years' time in part because of its form of government. Its highly centralized, one-party system more or less lacks meaningful checks and balances on executive power. Once the decision to pursue the massive subsidization scheme was made, there was no contrary legislative body to block it or slow it down. In contrast, representative democracy can be a long, raucous ride to policy outcomes that are often much watered down to satisfy multiple constituencies—a dynamic on full display during the fractured debate over the 2009 US stimulus bill, passed without a single Republican vote weeks after President Obama assumed office.

This point leads to a second observation: ultimately, China's takeover of solar in the wake of the financial crisis happened not only because it decided to prioritize the industry but also because it had no serious governmental competition, including from the United States. Europe and Japan, home of most of the erstwhile leading manufacturers, were both reeling in the wake of the crisis and were in no shape to respond to the tidal wave of Chinese support for solar. The United States was somewhat better off and did choose to pivot strongly to renewables under the Obama administration's stimulus plan but, critically, its approach not only presented no obstacle to China's gambit but rather *complemented* it. Like other key facts about solar's remarkable story over the past decade, this mutually beneficial dynamic between US and Chinese industrial policies at this time has gone largely overlooked.

China's green energy response to the financial crisis was to support solar in order to promote manufacturing, while the US stimulus response promoted getting renewable generation projects built. It wasn't that the US response did not prioritize manufacturing jobs—the $80 billion spent bailing out the automobile industry demonstrates this point—but rather that it did not prioritize solar manufacturing. China deployed its $47 billion in subsidies broadly and deeply across its solar manufacturing sector, rapidly injecting billions directly into manufacturing companies and extending them additional billions through loans, as well as directly supporting export deals. In contrast, the US stimulus prioritized rapid deployment of off-the-shelf equipment in construction of new projects, leaving the choice of whether to use equipment manufactured in the United States or elsewhere completely up to the company building the project—a feature, by the way, firmly in compliance with international trade norms. This kind of approach would only grow American wind or solar manufacturing if paired with a stimulus separately targeting these kinds of American manufacturers, something the stimulus package did not do. Of the Obama administration's $840 billion total stimulus passed by Congress in 2009, about $92 billion was committed to "clean energy technologies," although this catch-all category actually boiled down to just $21 billion for renewable energy.[2] Of this, the vast majority of funds went into tax credits, cash grants, and loan guarantees for projects, with relatively small sums dedicated to renewables manufacturing (and these mainly to wind, not solar).[3] An exhaustive 2014 assessment of the Obama stimulus package lists $1.1 billion in tax credits and four loan guarantees going to solar manufacturers, two of which later went bankrupt (including the infamous solar company highlighted by Mitt Romney's 2012 campaign, Solyndra).[4]

The biases in the US stimulus package, in part, reflect fractured American constituencies with regard to renewables and climate change. Consider the controversy following President Obama's deployment of financial crisis stimulus funds in 2009 to clean-energy purposes and the drama surrounding "picking winners" like Solyndra. This debate as to whether and how the US government should be involved in clean-energy and climate initiatives is even more contentious today, while Chinese guidance of its solar growth agenda has remained steady and will likely continue to be.[5] The Solyndra controversy can be seen alongside another instance where the US got deeply involved in the affairs of certain corporations, but without most of the same political drama: the auto industry. The $21 billion for renewables equals only about one quarter of the $80 billion spent to bail out the automobile industry, suggesting that politicians will agree about saving manufacturing jobs unless they are renewable energy manufacturing jobs.

A third observation about winner and losers in the solar development saga is that, while China has unquestionably created massive numbers of domestic solar manufacturing jobs through its solar policies, it is similarly undeniable that these same policies have created an explosion of solar-related jobs in importing countries. China has secured manufacturing jobs at home for itself totaling about 1.3 million.[6] However, the cheap panels that China exports in turn create jobs in the United States associated with the construction, installation, and operations of solar projects. Of the 242,000 solar jobs in the US, about half are in installation and construction. Someone has to develop, install, and maintain the panels at projects like Barilla and on rooftops across the country.[7] It may seem, looking at these figures, that China has gotten the better of us again, but given that China's population is much larger and younger than that of the United States, the weighted difference in overall solar job creation is broadly proportional.

On the question of whether China's solar policy has cost the US jobs, it was telling how many solar advocates in the United States—including the main solar industry trade group, the Solar Energy Industries Association (SEIA)—declined to support the Trump administration's imposition of tariffs on China for their solar trade practices. These groups have seemed to believe that, on balance, putting up with China's unfair practices is worthwhile, because continued access to cheap panels has become a very effective way to grow the US solar industry. As SEIA put it in a statement on the imposition of tariffs on Chinese solar panels in the news, "Although we strongly support U.S. manufacturing, SEIA has publicly taken the position against this petition to avoid damage to the 9,000 companies and 258,000 jobs in parts of the solar industry."[8] SEIA and many others saw the chances of competitive solar manufacturing returning to the US as low, so the trade dispute that the Trump administration conjured amounted to a threat to the many existing jobs in today's solar industry, which depend on a continued flow of cheap panels that need installing, operating, and maintaining. The incoming Biden administration has espoused strongly pro-solar policies and therefore seems unlikely to upset today's status quo, although it is worth noting that the Biden campaign also strongly emphasized increasing clean-energy manufacturing jobs lost to Chinese subsidies in language surprisingly reminiscent of his predecessor.[9]

A fourth observation is that none should forget that China carries a heavy price for its solar investment, both in debt and, perhaps surprisingly, pollution. These points are probably the least capable of being documented with exactness due to the murky information about both topics that is publicly available in China. It would be impossible, for instance, to estimate a cumulative figure for all the public and private debt extended to the industry, although what information is public suggests it is at least $47 billion saddling the Chinese economy. Whatever the total figure, there is no question that Chinese lenders have borne enormous risk for the country to cover so much ground so quickly in solar manufacturing.[10]

The same can be said for the troubling environmental footprint of the industry in China.[11] China has rich silicon reserves, but the processes to create polysilicon out of it result in many dangerous pollutants. One of these, silicon tetrachloride, is a highly toxic pollutant that transforms into hydrogen chloride gas in the open air. *The Washington Post* documented instances of Chinese silicon companies simply dumping silicon tetrachloride in the countryside in order to avoid the high cost of recycling or properly disposing of it. The *Post* also documented companies periodically burning off mysterious substances at plants.[12] A lack of strict environmental regulation and enforcement means that high levels of poisonous waste from solar manufacturers may be a problem for years to come in areas where it has been dumped.

In addition to industrial pollutants, many aspects of solar manufacturing are highly energy-intensive, meaning that significant Chinese coal power plant emissions have contributed to the carbon footprint of the clean solar panels exported around the world. For many Chinese citizens, the irony of making the world's cleanest energy source using power from some of the world's dirtiest coal power plants, so that other nations can preserve cleaner air for their own populations, is surely a bitter aspect of China's push into solar manufacturing.

A fifth observation is that while it is true today that China dominates the solar industry worldwide, the same may not be true tomorrow. China and the West each remain contenders for domination in "next-generation" solar technologies. China's unique advantages in deploying capital and subsidies into photovoltaics were transformative in large part because the industry today is fundamentally a scale manufacturing exercise. This is not to downplay the multitude of iterative improvements that Chinese companies have introduced to photovoltaic manufacturing processes but rather to emphasize that the basics of photovoltaic manufacturing were already established when China began its surge into dominance, while the next developments in solar could arise differently—for instance, from step-changes that move the technology in completely new directions. As secure as China's position is today in photovoltaic manufacturing, research and development into new solar technologies and applications is strong in many countries, and it would be difficult to guess which may be best positioned to dominate should a new type of solar technology emerge.

US companies have had a lead in thin-film solar technologies, a form of photovoltaic generation using a chemistry that is different from that of silicon-based approaches. But thin film still lends itself to economies-of-scale savings in much the same way that silicon cells do, so China's advantages could come to bear there as well. A different technology, perovskite crystals, is currently in development in labs in many Western countries. Perovskites are photovoltaic materials that hold the promise of potentially being sprayable in a paint format and may eventually be even cheaper to manufacture than silicon or thin-film cells. China may own the current solar manufacturing industry, but this does not guarantee that it will remain in this position as new solar technologies inevitably come into play.

■

Let me put forward one last way to assess the true winners and losers of China's transformation of solar manufacturing. This take on the matter involves putting China's massive investment to transform solar into perspective alongside similarly massive capital expenditures.

Let's start by looking at exactly how much China invested in total to create its solar dominance. Unfortunately, as mentioned above, it is not possible to calculate this amount with certainty due to the opaque nature of public sector data in China, but it can be estimated. Efforts have been made in trade dispute filings and academic inquiries to identify the total value, and the best figure available is approximately $47 billion (including grants, loans, and loan guarantees) over the period 2009 to 2012.[13] Given the diversity of all subsidies offered at various levels of government and from different departments, and the blurry line between state-affiliated banks and private ones, it seems likely that this figure would reflect a low estimate, perhaps low by an order of magnitude. Even so, $47 billion in 2012 dollars, or about $43 billion in 2017 dollars, is an extremely large sum.

It is difficult to place a figure like $40-odd billion in context, but perhaps the easiest way to do this is to compare it to other large projects as points of reference. For instance, Boston's "Big Dig," the infrastructure project to rearrange transportation in downtown Boston to and from the airport and other central locations, cost $24.6 billion in 2017 dollars. The construction of

Great Britain's Channel Tunnel cost $23.9 billion, and the largest infrastructure project in the world, China's Three Gorges Dam, cost $42.1 billion—also both in 2017 dollars.

Darker points of reference are also obtainable. The cost of the California wildfires of 2017 has been estimated at $19 billion, and the 2018 fires estimated at $24 billion; costs for the terrible 2020 fires will likely exceed these significantly.[14] Hurricane Sandy has been estimated as causing damage in the amount of $71 billion, and Hurricane Maria was estimated at $90 billion. It is intriguing to consider that, for half the funds that went to recovery efforts for a single hurricane, China may have funded the most potent weapon yet to combat the climate change that is blamed for super-charging some of the deadliest tropical storms in recent years.

My own point of reference is the damage caused by Hurricane Harvey, the unusual storm that struck Houston in August 2017. This storm came ashore on the Texas Coast south of Houston, but then drifted north and stalled over the greater Houston metropolitan area. Over a period of four days, more than five feet of rain fell in several places, breaking all US rainfall records except for one, a freak rain event that occurred in Hawaii. Parts of the city that had never flooded before were underwater and remained so for weeks as reservoirs were forced to release even more water to prevent worse disaster. Most of my family had moved from Galveston to Houston long ago, only to be flooded again in Houston, and we hosted some family members in Austin as they waited on the floodwaters to recede. My uncle's home, backed up to a Walmart parking lot that had become a vast lake, had almost two feet of water in it. The total damage caused by Harvey was estimated at a whopping $129 billion, more than four times what China may have spent to reinvent solar.[15]

Perhaps the most on-point framing of China's solar investment does not have to do with infrastructure or natural disaster precedents but rather the undertaking in science arguably most analogous in scale: the Manhattan Project, which all told cost the

United States about $25 billion, in 2017 dollars. It is incredible to imagine that the herculean effort that went into the development of the atomic bomb cost a little less than half of what subsidies to Chinese solar panel manufacturers cost, reinventing photovoltaics as an incredibly competitive power source over just a few years.

If you consider the funds China invested not as a rogue trade practices subsidization scheme but instead as China's investment in its own Manhattan Project—which it is now gifting to the world year in and year out in the form of massively subsidized solar panels—then it seems clear that a significant burden has been borne by the Chinese taxpayers who funded the endeavor, a burden far outweighed by the direct benefits they received.

Thus, there is a powerful argument that the net final effect of the massive Chinese subsidies was to convey to American and other energy consumers around the world, somewhat circuitously, the enduring benefit of Chinese industrial policy largesse in the form of ever-collapsing solar panel prices.

As strange as it may seem, from this perspective Chinese taxpayers paid to have Chinese workers produce too-cheap solar panels and then sell them to the US and elsewhere at a large discount relative to real market prices. Americans have ended up with cheap panels that will generate cheap electricity for decades, displacing emissions each step of the way and producing more positive climate, environmental, and human health benefits with each year that passes. The price for us to receive this bargain is the sacrifice of investments in US, European, and other companies, as well as the loss of jobs through which the same panels would have been manufactured (at much higher prices). These losses should not be minimized, nor should we minimize the fact that China violated its international trade agreements to cause these effects. China should be held accountable for the violations, and we must acknowledge that, for individuals who lost their jobs in these companies, the larger benefits may not exceed the costs. But while everyone is entitled to their own view of the personal net outcome of this arrangement, in the context of the climate realities we are facing, this aspect of the deal unquestionably creates very significant positives.

Insofar as photovoltaic solar today represents mankind's most promising tool by far to combat climate change developed to date, China's reinvention of the technology—even acknowledging the self-serving aspects of the policy's intent and implementation—amounts to an extraordinary gift to the world financed by Chinese taxpayers.

Especially during the Trump years, it was easy to ignore the positive side of the ledger when it comes to China's ruthless domination of solar manufacturing since 2006, and instead focus on the unfairness of China's trade practices and the jobs and investments lost by Western companies in the face of China's growing manufacturing prowess. In the end it will depend upon what the ledger tracks: if trade compliance matters most, then China's misdeeds are the takeaway, but in the carbon history of the world to be written 100 years from now, this detail fades entirely.

And herein lies perhaps the most important lesson that we can learn from the history of solar technology, from Becquerel to Bell Labs and from ARCO to Yingli. Its unique history shows that, possibly more than any other electricity-generating technology, solar is a child of the world, both in terms of where it came from and what it may achieve. The technology was born in France, became a teenager in the United States, went to college in Japan and Germany, and then went to China, where it became a global superstar. Today, its panels are likely to be found just about anywhere you go in the world, not only providing cheap energy but also advancing efforts to address climate change. Globalism presents many challenges, but in looking at all the facts of this long story, it seems clear that the advent of solar is an important symbol of the great achievements it makes us capable of: that when the world works together on something, with different nations contributing according to their strengths—even in an uncoordinated way—remarkable results are possible.

PART 3

FIXING OUR CAPITALISM TO TAKE ON CLIMATE CHANGE

CHAPTER 9

MOMENTS IN TIME

HOW OUR ENERGY HISTORY HAS CREATED OUR FUTURE

Putting our brief and glorious, but ultimately disastrous, affair with fossil fuels into a historical perspective.

Rachel Carson, perhaps the most effective voice on environmental matters the world has ever had, is best known for her visionary 1962 book *Silent Spring*, which starkly details the cumulative negative effects across the natural world of pesticide use. For me, however, her earlier book, *The Sea Around Us*, is her masterpiece.[1] This book is a fascinating tour through all the different strata of the world's oceans, as filled with facts, data, theories, and explanations as the landscapes she describes are filled with creatures and captivating phenomena. One doesn't even realize how densely packed with information each paragraph is, so lovingly and compellingly is it written and, more importantly, so obvious is the passion and zeal of the author for her subject. By the end, one comes to appreciate more than anything the supple mind of the author—its command of a quantum of information vaster than most other scientists, but tethered so intimately to her personal connection to the sea that it hardly seems scientific at all.

Given the scope of *The Sea Around Us*, Carson's brilliant mind and careful eye, and the significance of the topic—the ocean, the element of our planet fated to experience climate change most profoundly—it is intriguing to read it now in light of what we know to be true. The warming trends so voluminously documented by scientists today did not elude her even then, nearly seventy-five years ago. "Now in our own lifetime," she wrote in 1951, "we are witnessing a startling alteration of climate.... It is now established beyond question that a definite change in the Arctic climate set in about 1900, that it became astonishingly marked about 1930, and that it is now spreading into sub-arctic and temperate regions. The frigid top of the world is very clearly warming up" (174). She observed glaciers increasingly melting, cold water fish such as cod migrating north, and major currents like the Gulf Stream warming and evolving. Carson struggled to find a reasonable explanation for what she observed, settling on an odd theory about deep water tidal flows, but she unmistakably documented warming patterns predating even her own life and a cascade of associated changes coming with it.

The Sea Around Us is not, of course, a book about climate change, and Carson's discussion of these warming trends is but a small part of one single chapter. She wasn't looking for the unusual things she noticed; she came upon them by accident, something amiss in the ocean worlds she otherwise knew so well. And in this one particular thing, we all share something important with her: we too have been living through climate change, mostly unknowingly, for our whole lives, as it increasingly intrudes.

For myself, I have seen the waterline at the Galveston beaches I grew up on creep up each time I return for a visit, the summer high temperatures rise each year at our home in Austin, Texas, and the winter lows climb each year in Portland, Maine, where my father was born and raised—although for most of those years I had no real explanation for what accounted for the changes. What Carson's accidental discovery reveals to us is that climate change has been going on each day of our own lives in the background, in ways much more personal to each of us than scientific, even as the phenomenon's larger outline, causes, and evolving effects remain deeply complex and inscrutable to us.

As I wrote in chapter 1, gaining an understanding of complex things like climate change is often a struggle for perspective, and this has been very true for me. My first understanding of climate change was mainly as a current event, a purely modern phenomenon, something like Styrofoam, recycling, or the Internet—something recently on the scene and just another challenge of modern life that we are sure to manage. In reality, though, the more one pulls on the thread of the greenhouse-gas emissions that we humans have released over time, and the buildup of sufficient quantities of them to influence the all-but-imperturbable balance of atmospheric composition formed over geologic time, the more one realizes that this is a much, much bigger story than Styrofoam.

I have pulled on this thread, and pulled and pulled some more, and have come to understand it as a phenomenon that began much, much earlier, and that will all but certainly continue for much longer than the timescales we are used to. The more I thought about it, the more I came to view the story of climate change as really being the story of people and energy. For us, a species afflicted with an endless resourcefulness and a propensity to dream, this story was certain from the beginning to be transformative to us and our environment. In truth, the broad gap between mankind's ability to imagine new things to create and do, on the one hand, and our limited personal capacity to get them done on the other, has been the void filled by the multitude of energy sources we derived and exploited, all the way from wood in ancient times to the ocean of combustion-driven, fossil fuel energy consumed each day today. From this perspective, from the moment we discovered the potent and compact bounty of fossil fuels, climate change became simply the predictable, even inevitable, byproduct of our ability to dream, plan, and execute the soaring achievements we have attained over the last few centuries.

Taking the long view, it is apparent that humankind's relationship with energy has been a deep, complex, and highly beneficial connection spanning the entirety of human history. Nearly indispensable from the very beginning of our species, this relationship has evolved in myriad ways, changing and growing, and changing us, in our spectacular climb to the top of the heap of life on earth. Almost without fail, energy has been at our side in support of human progress through our many struggles, overwhelmingly a force for good, for work, for advancement and improvement. It has allowed us to protect more, feed more, clothe more, and comfort more, even as there were more and more of us to protect, feed, clothe, and comfort. In modern times, our mastery of energy and its propagation into machines, devices, and gadgets has made it an omnipresent actor deeply integrated into our daily lives.

For most of us, energy and what we make from it manifest so seamlessly and intimately in our affairs as to be transformative to our very natures: an extension of our minds and bodies beyond our physical selves, a means of hive-like connection between us wherever we may be, a protective envelope surrounding us no matter how inhospitable the environments into which we venture. In the twenty-first century, energy permeates most of human existence to such an extent that to say it is taken for granted would be a grave understatement; imagining its complete absence is as unnatural to us as finding ourselves in a pitch-dark room and not reaching to turn on a light or strike a match. The pages you hold before you, the glowing screen within arm's reach or across the room, the clothing and the heated or chilled air surrounding you, the coffee or tea in your cup, the water you bathed or showered in—all are steeped in energy in ways we give almost no thought to. If the energy put into everything in our world was visible as a color, it would saturate almost everything we touch and see as we go through our days.

It is a cruel paradox, then, that through climate change, the same indispensable resource fueling our daily existence and all our greatest achievements has now become the greatest threat to them. How ironic—or perhaps not, considering how climate change resulting from greenhouse gas emissions was foreseen as early as the nineteenth century by scientists such as Eunice Foote and Svante Arrhenius[2] and has been general knowledge for more than three decades now—that at the moment of the apex of our dependence, this relationship is turning the world we created with it against us.

Consider the example of a not-so-innocent tomato sitting on your kitchen table. The delicate fruit, originally native only to South America, is grown and enjoyed today almost everywhere on the planet. The United States produces about 11 million tons per year, Europe about 15 million tons, and China almost 60 million.[3] It may seem at first glance that the tomato would require practically no energy to make its way to your table—surely it just grows in a field until it is picked and driven to your grocery store, where you bought it and carried it home. In fact, there is much more to the story.

A good deal of tomato production worldwide is actually done in greenhouses requiring large amounts of electricity to heat and/or cool, sometimes year-round. Water for irrigation must be provided, production of which also requires significant amounts of electricity, particularly in farming areas remote from water such as California's Central Valley. Like every other popular crop, tomatoes also require significant amounts of fertilizer to grow, and the manufacturing of fertilizer is an extraordinarily energy-intensive process. The Haber process for creating powerful nitrogen fertilizers was invented in 1913, revolutionizing farming and paving the way for the industrial agriculture system we know today. To make the fertilizer your tomato needed to grow required burning natural gas to heat yet more natural gas to about 400°C.[4] The resulting ammonia compound delivers the nitrogen needed to powerfully enhance plant growth when spread on cultivated lands, but applying it also releases nitrogen oxide, a greenhouse gas hundreds of times more potent at trapping heat than CO_2, into the air.

When the tomato is finally ready for harvesting, agricultural equipment, probably powered by diesel fuel or more electricity, was involved, and then it had to be trucked to market. Most tomatoes in the US are grown in California or Florida, so depending on where you live it may have been driven across several states to your store. All in all, scientists estimate, the processes required by the tomato industry produce so many emissions that on average each and every tomato has a carbon footprint in greenhouse gases more than twice its weight.[5]

This simple but powerful example shows how even things we tend to think of as wholesome, natural, and good for us can be, in fact, significant sources of greenhouse gas emissions. Water is another innocuous-seeming example. What could be more natural and free of carbon pollution than a simple glass of cool water? It turns out that the energy required to produce the water we drink and use every day is massive: extracting it, treating it, pumping it, transporting it, and so on, all day, every day, all year, in every part of the country. Water-related energy use in the United States is equivalent to about 13 percent of the total amount of energy consumed in the United States, an amount about equal to sixty-two coal-fired power plants, and about 5 percent of our total greenhouse gas emissions.[6]

These modern processes are causing the earth's environment, which birthed and cradled us from our ancient *Homo sapiens* beginnings, to be transformed in real time into a substantially more hostile place. Increasingly, changes to our environment are coming to be governed less as we intend by the laws of man and more as we do not by those of chemistry, physics, and math. Crucial atmospheric and oceanic chemical balances are drifting out of kilter, glaciers and ice sheets are cracking and melting, temperatures and sea levels are rising, rainfall amounts are diverging wildly from norms—changes we all might read about in any newspaper today. Barring some transformative positive development—which, to date, there has been little reason to expect—we, the cleverest and seemingly most favored of species, seem poised for a dramatic fall that will be as tragically foreseen as it was preventable.

Into this picture comes the story of the small solar plant that is the subject of this book. This dramatic and promising new moment in which solar has begun to emerge as the new sheriff in town, as it turns out with the benefit of the long view of history, does not exist in a vacuum. In fact, it is a third moment in the human energy story, one that cannot be properly understood without first looking back at the prior two. Taken altogether, these three transformative energy moments in our history trace the trajectory of our journey up to today and mark the departure point for starkly divergent potential futures.

■

For the first of these moments, it is necessary to go back in time—way, way back to nearly the very beginning of us, to a moment somewhere between one and two million years ago when, archeologists surmise, one of our particularly observant *Homo erectus* ancestors first noticed that sparks were produced when one rock was struck against another. When she did it again, teasing the sparks into flames, it meant that the first and only species of the billions that have existed on earth had deliberately conjured fire, claiming the power to make it appear at will.

This single extraordinary moment launched a series of entwined events over human history that continue to this very day. Fire, a chemically simple oxidation process similar to rust but infinitely faster, works through the rapid combination of oxygen molecules with other elements to produce two of what were surely the rarest and most treasured of commodities in the daily life of *Homo erectus*: heat and light. The reliable creation of fire must have eventually led to its mastery as a blunt but multifaceted tool and, in turn, to its enduring use. One can imagine an unbroken line extending across these million or two years in which humans came to rely fundamentally upon fire, first simply to banish cold and dark, then to cook food and defend ourselves from predators, and then to a constellation of diverse applications.

The fuel sources used by *Homo erectus* all the way through to the blossoming of our *Homo sapiens* species, about 300,000 years ago, were almost exclusively things that naturally renewed themselves: primarily wood, but also, to a lesser extent, straw and dung. We know this from the archeological record, which abounds with firepits, appearing with ever greater frequency as the millennia advanced. Today we would call these resources "biomass fuels." They were the first renewables, in the sense that whatever was burned was more than amply replaced by natural processes such as the thriving old-world forests, which absorbed more or less all the carbon dioxide that burning them released. Wood and straw grew plentifully enough through the glacial cycles of the Pleistocene to ensure that humans had plenty of fuel, even through the end of the last Ice Age, only about 11,000 years ago.

When the last glaciers finally retreated, ushering in the Holocene—the current epoch of geological time—agriculture suddenly emerged, and human populations grew and varied their energy mix significantly. One historian calls the Holocene the "10,000 warm years," which enabled the explosion of human civilizations that have culminated in the modern world, noting that this anomalous period enjoyed average temperatures between 1° and 3°C (1.8 and 5.4 degrees F) warmer than preceding glacial periods.[7] We became less migratory at this time as we tended our crops, and in doing so we developed relationships with certain animal species like horses, oxen, and camels, domesticated them, and expanded our energy portfolio to include animal muscle. We also developed new ideas about how to improve our lives with the sudden free time that specialization allowed, ideas that often required even more energy.

We began to rely more than ever on fire as we came to use it for a wider variety of purposes: to clear farmland, wage war, cure skins for clothing, create ceramics and bricks, experiment with metal-lurgy, forge tools and weapons, and much more. In many parts of the world, the tragedy of enslavement and conscription began around this time, which added concentrated human muscle to the energy mix—a fact darkly memorialized in many enduring monuments of antiquity and extending well into modern times. We also discovered how to tap other energy sources, such as water and wind, to do our work and to power new inventions. Throughout the early Holocene, these new power sources were deployed around the globe to power the growing portfolio of human inventions for an ever-expanding list of uses: milling grains, cutting stone, preparing materials to make textiles, making paper and wire, and on and on.

Powered by these widely deployed energy sources, early human civilizations thrived, grew, and spread across the globe, including into areas previously too inhospitable, distant, or hostile. For much of this time humans were beset by threats—not only larger and stronger predators but also disease, starvation, and drought—all challenges they found new ways to outsmart and overcome. With agriculture providing a more stable and increasing food supply relative to what was available in the hunter-gatherer era, and with increasingly potent energy sources at hand to extend humans' reach to execute their wishes, the stage was set for something new. It was at this point that complex human societies appeared in many areas around the globe, more or less simultaneously in Asia, the Middle East, the Americas, and Europe. These early civilizations, in turn, spread across the world, evolving over time to create the world in which we live today.

To a hypothetical visitor from space looking down on the earth's life forms 100,000 years ago, our persistence and triumphs would likely have seemed surprising. We *Homo sapiens* are relatively puny, nearly hairless creatures who can't run very fast, who bleed easily, and who bear young utterly helpless long after their birth. This is hardly the description of a creature suited to mastering life in the Stone Age. But our canny intelligence and resourcefulness, progressively amplified by our persistent devotion to mastery of an ever-widening list of energy sources, endowed us with myriad advantages that helped us to prevail, prosper, and thrive. We began the "10,000 warm years" period numbering only about 2.4 million across the globe, and by the beginning of what historians call modern history, around AD 1500, we were more than 450 million.[8]

It is easy to overlook an important aspect in this remarkable saga of early mankind. From the discovery of fire until very recent times, a period of more than a million highly prolific years, our species' dramatic expansion was underwritten by numerous energy sources, all of which were renewable. If one were tracing the energy history of humanity, it would be notable that up until a very short time ago, in geological terms, humankind's early technologies were almost exclusively fueled by energy sources that naturally renewed themselves. The use of these sources had, by definition, practically no negative consequences for the climate or the environment. For better and for worse, this was about to change—and along with it, just about everything else.

■

Fast-forward a few hundred years to the beginning of the eighteenth century, when the second important moment in the story of humans and energy occurred. A series of advances that together matched the discovery of fire in significance led to the worldwide deployment of new, denser, and much more potent energy sources.

Prior to 1700, coal and oil had been discovered and burned for various purposes, but they had never been used in the systematic and widespread manner that biomass fuels had been.[9] Around this time, however, it was gradually recognized that coal and its slightly refined, less odiferous cousin, coke, both burned hotter and lasted longer than wood and other biofuels. They were also much cheaper and more plentiful, if you knew where to find them. It is perhaps no coincidence that this awareness occurred in England just as exciting new technological applications began to emerge that were very well suited for hotter-burning and longer-lasting fuels—roasting of malt for beer, for instance.

These developments together led, in 1709, to coal's first use as a preferred replacement for wood and charcoal in blast furnaces used to make pig iron, a very crude form of the metal. The first rudimentary coal-fired, steam-driven machines began to emerge around this time, not just as technological innovations but also in response to the social and economic trends of the day that were driving new business opportunities. The confluence of new technologies, denser and more energetic fuels, and these new economic opportunities led, of course, to the early days of what we call the Industrial Revolution, representing nothing less than the stirrings of modern capitalism itself.

The first steam engine—as we will see, the mother of all inventions when it comes to combustive technologies—appeared in 1712. The steam engine was created by Thomas Newcomen, a lay preacher moonlighting as an inventor outside of London. His crude device was based on somewhat similar predecessors but was sufficiently different to be widely acknowledged today as the core technology behind a machine that went on, in every sense of the phrase, to conquer the world.[10] This domination continues to this day and probably will for a significant part of the foreseeable future. It is easy to imagine that the steam engine's progeny may never disappear, so useful have these machines been to humans in so many settings across the modern world. One might even argue that no expression of human ingenuity has ever been propagated more widely, persistently, and effectively in so many places across the planet, and with more consequence, than Newcomen's engine and its innumerable offspring.

Given the state of science and technology at the time, Newcomen's crude steam engine was a breathtaking achievement. It was a massive device, housing a boiler, piston, and other equipment inside a brick structure about forty feet in length, from which emerged a gargantuan twenty-five-foot rocker beam connected to the pump mechanism. A vacuum created by condensed steam drove the piston, which, once primed, raised and lowered the beam at a rate of about twelve times per minute. The engine's chamber was filled with steam and then cooled with a spray of cold water, causing the resulting vacuum to draw the rocker beam back down, thereby operating its pump.[11] One of the steam engine's first deployments was, ironically, to pump out coal mines that had become flooded by seeping water, thus clearing the way for more coal to be extracted (a coal-fueled motor used to enable more coal to be mined, necessitating more pumping, to remove more coal, in a great loop). Initially, other applications for the machine were few and far between—Newcomen's device was so large, expensive, and inefficient (operating at just 1 percent efficiency) that its prodigious fuel requirements kept it largely tethered to the mines that fueled it.[12]

Conceptually, Newcomen's discovery was how to convert the potential chemical energy in coal into kinetic energy to get a job done that was otherwise all but impossible using labor and the other technologies of the day. His idea—burning coal to convert water to steam, and then manipulating the gaseous properties of steam to drive the work—was the innovation, one that has been relentlessly and incessantly refined, revised, improved, and revamped in the intervening centuries. Inventors soon discovered that the expansionary force of water becoming steam was superior to the vacuum effect of condensing steam, and in time they created many ways to harness the expansionary effects of steam, and later of other fuels, much more efficiently. Although it would be wholly unrecognizable to Newcomen and his fellow early innovators, the principles of Newcomen's engine are at the center of the engine in the car sitting in your driveway and the fossil fuel power generators and plants found everywhere today.

Very slowly at first, but steadily and at an ever-growing rate, the power unleashed by Newcomen's new technology increased, coming to be measured not by the equivalent strength of men but by that of horses, and then soon in tens or even hundreds of horses. In the late 1780s, James Watt redesigned Newcomen's engine and made it 75 percent more efficient, and then with his partner, Matthew Boulton, further improved it by adding a turning motion for doing work, as opposed to the simple up-and-down motion of Newcomen's version—an innovation that greatly increased potential applications. Throughout the eighteenth and nineteenth centuries, the common element of a vast number of new machines was parleying the unusually dense and easily transported energy found in newly available fossil fuels into modern conveniences.

At the beginning of the nineteenth century, two new important applications materialized. Richard Trevithick invented the steam-powered locomotive in Cornwall, England, based on his study of the numerous Newcomen engines installed to pump out Cornish mines. And across the Atlantic, Robert Fulton invented the steamship, a development that in very little time resulted in both the figurative and literal remapping of the world. Then in the 1880s, another world-changing new technology, the coal-powered electric generator, began to create electricity for light bulbs and, in time, all kinds of other devices. Technology to generate electricity had of course been around since the 1820s when Michael Faraday's crude motors transformed electrical energy into mechanical energy, but as late as the 1870s the spinning work in dynamos was being done by horse-drawn treadmills[13] or coal generators inherently limited by size and efficiency, such as that invented by Thomas Edison.[14] The relatively compact generator invented by Sir Charles Parsons in 1884 used coal-fired steam passing through a turbine to rotate a multitude of copper wires through a magnetic field, creating a flow of electric current.[15] Parsons licensed his generator to wealthy American entrepreneur George Westinghouse, and it and competing technologies swiftly spread across the United States, then through Europe, and then, slowly but surely, to just about everywhere else in the world. Within two centuries of Newcomen's discovery, coal-powered engines in mines, trains, ships, and power stations had become an irresistible and ubiquitous resource deployed around the globe in civilization's service.

In time, coal was joined as a fuel source for these machines by oil and then natural gas, and all the respective chemical cousins and permutations of each. In this way, slowly at first but soon with an overwhelming momentum, fossil fuels came to more or less replace their renewable predecessors as fuel sources for most of the world. Oil had been used in many parts of the ancient world for everything from street paving to lamp fuel, but it wasn't until refining processes were discovered—in particular, refinement of kerosene in 1850—that large-scale usage began. Natural gas, originally a nuisance substance to be managed before oil could be obtained from a well, also came to be valued as a fuel in its own right. It was used first as a lighting source, and then for myriad applications by the middle of the twentieth century. New machines like combustion engines, distant relatives of Newcomen's engine but still clearly in the family tree, took advantage of the highly ignitable nature of refined oil products to do away with steam entirely, instead relying on the combustion of the fuel itself for the expansive force.

The many advantages of these fossil fuels were hard to miss—they were cheap and hundreds of times denser and more portable than biomass—while their biggest shortcomings, mainly the paralyzing smog they produced initially and, later, the invisible but enduring CO_2 pollution and geopolitical conflict associated with their extraction and distribution, seemed like small prices to pay at the time.

In this way, Newcomen's coal-powered steam engine catalyzed a powerful cascade of developments. It begat the coal-fired steamship, the railroad, and the power plant's steam turbine, which begat the automobile's gasoline combustion engine, which begat today's combined cycle natural gas turbine, the jet engine, and even the space shuttle propulsion rockets. Jet- and rocket-propulsion systems are highly complex interactive systems, but the essential principle at their core echoes the same central concept: the combustion of a fuel source resulting in burning gases expanding powerfully out of compressed spaces, causing an acceleration force used to do work.

The burgeoning diversity of combustion-based machines was matched only by the scale at which we propagated them to every corner of the globe, a scale advancing at a rate difficult for the human mind to comprehend—a geometric rate of increase so great that its best analog may be bacterial reproduction. Today, you are likely within range of some version of this technology whether you are standing on a street in Chicago, Lagos, Lahore, or Ho Chi Minh City; visiting a farm in Kentucky, a Russian steppe, a Kenyan savannah, or China's Manchurian plain[16]; or if you're sitting in a plane, train, bus, or ship somewhere in between. Combustion engines are everywhere you turn, and their descendant technologies—most notably, electricity—are virtually omnipresent in human population centers. There is no country on earth that does not use steam-to-electricity technology powered by fossil fuels, a fact that helps to account for the fact that nearly 90 percent of our species today has regular access to electricity.[17] In the vast majority of those parts of the planet where humans are present in significant numbers, there are few aspects of a person's life that do not rely in some form or fashion on fossil fuel electricity.

I first learned about most of these major technological advances—the invention of the steam engine, the steamship, the internal combustion engine, etc.—in my fifth-grade history class in Galveston in 1979. I was captain of my school's team in our citywide Quiz Bowl competition, where we tried to memorize a packet of about a hundred pages filled with all kinds of facts. The nation's bicentennial had been celebrated a few years earlier, and the context in which I learned about these various inventions was a source of pride about our prodigious ability to advance technologies and America's unique industriousness, ingenuity, and capitalistic aplomb. I understood these giant leaps forward were the things that made us great, the things that distinguished us as a country and constituted the steps along the way to our meteoric and inevitable rise to world economic superpower status. Now, in retrospect, we can see them as these same things, but also something very different: the giant leaps that paved the way to the grave challenges we face today.

■

These two epic species-altering moments—first, discovery of fire and its growth as a ubiquitous tool of early human ingenuity, and second, the invention of the steam engine and its fossil fuel-driven progeny—have defined the arc of energy in the human story thus far. For the entirety of *Homo sapiens'* 300,000-year history, we survived and then thrived by relying heavily on the energy sources available to us, the utilization of which were as integral to our successes as any other factor you might name.

As mentioned above, from an energy history point of view it is worth noting that during this 300,000-year period, humans relied almost completely on renewable fuels for all but the past 320 years or so. This means that, using round numbers, we spent approximately 299,780 years, or 99.9 percent of our species' time on earth, fueled completely by renewables, and only the tiny remainder fueled by fossil fuels.

It is a bitter irony that this geologically minuscule dalliance with fossil fuels is predicted to have such grave consequences. During those first 299,780 years, the carbon released into the atmosphere by burning wood and other biomass fuels was roughly equivalent to the carbon that new plants absorbed as they grew, a cycle of release and absorption that left mostly undisturbed Earth's natural atmospheric balance. What has happened during the past 320 years, of course, has been a substantial disruption to this balance.

As anyone who has seen *An Inconvenient Truth*, read a newspaper or magazine recently, or even just perused their Twitter or Facebook feed this morning would know, this disruption is what we have come to call "climate change."

Let me take a small detour here to offer an analogy that might help elucidate these proportions. Mathematically, this 99.9 percent figure is more or less the same as the time you would have been faithful to your college girlfriend or boyfriend had your relationship survived all of freshman year, save for a single night (in this case you would have been 364/365ths, or 99.7 percent, faithful). Humans have been extremely faithful to their first love—renewable energy—except for this single one-night stand with fossil fuels. This affair has truly been as brief as the blink of an eye relative to the 300,000-year timescale of humans' entire existence. But sadly, perhaps just as it may have been with your college significant other, that brief infidelity turns out to have made a huge difference. It may have been a thrilling and enjoyable evening, but the morning after, the cold light of day revealed the consequences.

The energy we have harvested from fossil fuels since the Industrial Revolution began has been the foundation upon which the whole of modern civilization has been built, including the multitude of dizzying human achievements our ancestors could not have even dreamt of. These advances have brought many positive changes, such as electrification, which has dramatically improved the quality of life for people around the globe. The more limited renewable fuel sources we used until around 1700 took us a long way, but our prospects changed fundamentally once we discovered how to access the vastly more energy-dense resources that had always been waiting just beneath our feet.

Growth in worldwide population and productivity can be traced directly to the changes that access to fossil fuels enabled. For most of the past 10,000 years, our numbers grew steadily but incrementally, starting at somewhere around 4 million and increasing to about 250 million during the time of Jesus Christ.

At the start of the Industrial Revolution in 1700, this figure had burgeoned to about 682 million souls worldwide, just about 100 million more than in the prior century. By 1800, the growth rate began to climb sharply: worldwide population increased by 300 million, topping 1 billion, and by 1900 it had increased by another 600 million.

This trend tremendously accelerated in the twentieth century. From 1900 to 1950, fossil fuels penetrated deeply into the world's major economies, intensifying economic activity and raising living standards which both, in turn, stimulated greater demand for electricity. This development in part facilitated the "Green Revolution" in agriculture, wherein irrigation and intensive fertilizers greatly increased crop yields. As these changes unfolded over the first half of the century, global population mushroomed by about 900 million, reaching 2.5 billion. Access to electricity worldwide hit 50 percent in 1970, and then 60 percent in 1990 as the population more than doubled to 5.3 billion.[18] Today, just a few decades later, access to electricity has reached almost 90 percent of the human population, while the population itself now tops 7.5 billion. Current projections foresee us reaching 9 billion—nearly double the 1990 figure—by 2040.[19]

If you find this blizzard of numbers unhelpful, focus on just a single fact: it took about 9,800 years for the modern human population to grow from 5 million in 8,000 BC to 1 billion in 1800, but in the mere 221 years since 1800—not coincidentally, the 221 years in which we discovered how to exploit fossil fuels at a furious rate—global population has grown by nearly 7 billion more.[20]

That population and productivity spiked just as we learned to utilize fossil fuels on a massive scale and began the Industrial Revolution is no coincidence. These trends reflect the fact that we "supercharged" our species' growth once we found the means to access—in addition to the energy that shows up every day from the sun—the unfathomable quantity of solar energy accumulated over billions of prior days and stored beneath us, concentrated in the form of coal, oil, and gas.[21] It is, of course, the plants that were nourished by sunlight over the billions of years before us, and the animals that ate those plants, that were the progenitors of the fossil fuels that we have become so adept at harvesting. It may be counterintuitive to think of fossil fuels as simply a chemical form of solar energy, but it shouldn't be. The only energy that exists on the planet Earth is sunlight, nuclear fuels, tides, and geothermal energy.

As dramatic as they are, the numbers describing the growth in population and GDP lately tell only part of the story about how fossil fuels have affected human existence since 1700. Even more important have been the many improvements in living standards catalyzed by advances in energy use. As a species, we have benefited from our persistent and widespread exploitation of fossil fuels in ways too numerous to count. In particular, coal—reviled today as the primary emissions villain—should rightfully be credited with making possible the electrification of cities and towns and the associated vast improvement in quality of life across the United States during the nineteenth and twentieth centuries, a boon also experienced by every other developed nation during roughly the same period. Coal also deserves significant credit for the United States' emergence as a major economic power during the nine-teenth century, fueling our greatest industries, including steel, railroads, manufacturing, shipping, and so much more. It could not be disputed either that coal underwrote US dominance over the next hundred years as well, during the "American Century"—including the Pax Americana that the world has experienced, start-ing in 1945 and extending, one hopes, for centuries more.

The desire of India, China, Turkey, Indonesia, and many other developing countries to achieve the same kind of results as the United States is the reason that coal plants continue to be deployed with regularity and why they will almost certainly continue to be for some time. China and India together accounted for 86 per-cent of all coal plants built worldwide between 2006 and 2016 and commenced construction of more than 40 gigawatts of new coal power plants in 2016,[22] about the equivalent of all the power plants operating today in New Jersey.[23] Even as signatory countries to the Paris Agreement have begun to comply with its stipulations, more coal power plants are being constructed. China's announcement, early in 2017, that it was terminating the construction of 103 coal plants—some of which were already under way—while also dramatically increasing cleaner energy generation was an important sign that the country is taking real steps to improve its overall emissions situation.[24] However, even with these cuts, China has continued to increase its coal use right through the COVID pandemic, building 38 gigawatts—about half a Texas in terms of electricity demand—of new coal plants during 2020, and its current five-year plan calls for adding more.[25] India, for its part, has also begun moving quickly toward renewable energy sources, but still, its overall emissions from coal power plants are projected to increase by about 90 percent by 2030, while Indonesia's will double and Turkey's will quadruple.[26]

Coal use is growing in many other countries around the globe, too. China continues to finance construction of large coal plants in power-hungry, less-developed countries, not only as investments but also to advance its geopolitical objectives. In Pakistan, for example—a country at the heart of China's new "One Belt and One Road" regional development corridor—China has announced financing and other support for approximately 12 gigawatts of new coal generation in the country, an amount equivalent to all the power plants operating today in the state of Maryland.[27]

It is not hard to understand the myriad benefits that these massive new investments in coal power plants bring to places lacking them before. Most of us in the United States take our reliable electricity service for granted—blackouts are extremely rare occurrences virtually everywhere in the country—but in much of the world, inadequate and unreliable power supply is not uncommon.

I have experienced this firsthand while traveling in places like Istanbul, Johannesburg, Marrakesh, and Delhi, where power was fairly consistently cut from time to time, especially during periods of high demand. When this happened, I would uniformly hear, a few moments after the lights went out, the roar of backup generators starting up all through the nearby neighborhoods. The sound was unbelievably loud, but much worse was the dark cloud of exhaust that then drifted up into the sky. The backup generators were diesel-powered and emitted a coarse black smoke all over town to keep the businesses running and households lit and cool or warm, depending on the season. During a severe outage, the generators would hum along for hours until finally every last one went silent, having emptied their diesel tanks. At this point, the city is transported back in time to the pre-electrification era, bringing to a stop any economic activity needing a light, computer, or machine.

When a giant new coal plant gets constructed, these kinds of problems disappear, replaced by a steady and reliable energy supply that families, businesses, and industry can all depend upon. Economic growth and improvements in quality of life for billions in developing countries today are intrinsically linked to access to plentiful and cheap electricity, which for most of modern history has been the unquestioned domain of large coal power plants. Increased availability of electricity represents a massive, tangible, and positive change in the daily lives of people previously living without it, or without sufficient amounts of it. Access to electricity also leads to other benefits, such as better economic opportunities, cleaner water and food, better health care and access to it, and, generally, lower mortality rates and greater leisure time. That these improvements all come at the cost of negative consequences to be visited upon future generations through climate change does not diminish how beneficial they have been and are every day. Speaking plainly about both the negative and positive consequences of our energy choices is necessary to clearly see the full picture before us.

We now know that the exploitation of fossil fuels, initiated at our second energy inflection point, has brought enormous benefits but comes with an even steeper price that we and our children and grandchildren will pay. We already know the outlines of this price, the details of which are becoming clearer in real time. Energy derived from these fossil fuels is at the heart of global emissions, both in its own right and in its role as the force behind myriad other emissions contributors, and thus to climate change.

These contributors include transportation, agriculture (including fertilizer production), and innumerable Industrial and Information Age practices. For instance, the Haber process, the chemical procedure described above for manufacturing the highly effective fertilizers essential to feeding the nearly eight billion humans presently alive, and cement fabrication, so crucial for the construction of our streets, buildings, and other infrastructure, are both responsible for massive quantities of greenhouse gas emissions each year. Information technology—including the data centers housing the servers that hold our emails and photographs, as well as the movies and music we stream on Amazon and HBO—are estimated to generate about 2 percent of global carbon emissions and have the fastest-growing carbon emissions of any economic sector.[28] All these uses make the energy sector the main event when it comes to climate change. It alone is responsible for 68 percent of annual worldwide emissions.[29]

And these factors account for only known and estimated emissions; other emissions from energy-related activities also play significant roles in changing the climate. For instance, we understand now that the natural gas infrastructure crisscrossing the United States could be leaking significantly higher amounts of methane into the atmosphere than previously thought.[30] Because methane has at least twenty-eight to thirty-six times the heat-trapping effects of carbon dioxide, any so-called "fugitive" methane emissions outside of current estimates would make things even worse than they already are.[31]

It is clear that our relationship with energy is getting turned on its head. As we seek to expand electrification beyond even today's massive deployment in order to lift the living standards of hundreds of millions more around the globe, that relationship is now poised, through climate change, to become in time a powerful force operating in the opposite direction. These many new fossil fuel power plants that we are building today are all piling new CO_2 emissions on top of the 560 gigatons we have put into the atmosphere over the past three hundred years of unrelenting expansion.

For perspective, a single metric ton weighs about half as much as a car, and a gigaton is equal to one billion tons. Thus by way of analogy, mankind has released a quantity of CO_2 into the atmosphere weighing the equivalent of about 280 billion cars, or about four cars' worth for every person alive today. Paradoxically, then, these new emissions have secured not just an improvement in quality of life but a significant diminishment to it as well, differentiated only by time and place: the factors causing enhancement have disproportionately improved conditions in wealthier countries and the diminishment will disproportionately affect the world's vulnerable populations.

■

We are not, of course, idly standing by as the challenges of climate change become fully apparent and the consequences begin to manifest. The same ingenuity that caused humans to notice and then master the spark that could start a fire, and that allowed us to harness steam to drive a piston, is increasingly being deployed to address this problem, and there is reason for hope. The prodigious track record of this ingenuity in making us a remarkably adaptable species should not be underestimated.

Change has never been so vital. The latest definitive report on the current state of climate change, authored by the UN IPCC, presents its forecasted outcomes along a spectrum of potential future emissions scenarios. The report classifies the consequences of lower emissions scenarios as being much more manageable than others. Sadly, there is no scenario in which we avoid harsh outcomes—very significant increases in temperature, sea level and ocean warming, and acidification are already "baked in" for centuries, even if we succeed in meeting the most aggressive targets yet agreed upon. But to the extent that we can realize the less grim scenarios, we will dramatically diminish the scale of risk we are undertaking—and thus, the alacrity with which we act now will make all the difference.

We are seeing changes today in the long human relationship with energy that represent the beginning of a turning away from fossil fuels and the hole we've dug for ourselves with them, and back toward our first love: renewable energy. All signs indicate that we must return to a path in which we bring our energy resources back into balance with our planet's means to recover from their exploitation. The economic momentum of the electricity and other industries behind our colossal annual emissions is enormous, but a growing number of innovators are finding ways to meet the same needs with far fewer harmful effects and are gaining an increasingly disruptive momentum of their own. Whether these ventures will suffice to effect real change quickly enough is uncertain. But it may be surprising to many to learn how rapidly we are turning in this direction.

If some of these innovators were to deliver a means of producing vast amounts of cheap energy that was 100 percent clean, electricity markets would change from being a barrier to reducing emissions into the most potent tool for cutting them. If renewables were to outcompete fossil fuel plants on their home turf—that of competitive power markets—fossil fuel plants would be starved of revenue, would diminish in value, and perhaps would even exit markets completely. In this way, market forces could cause the greening of our electricity industry far more quickly than government fiat would ever be capable of doing, and, even more importantly, they would do it without awaiting a political solution that seems far off. Furthermore, such a change could provide the basis for the other crucial ingredient needed to prevent the worst effects of climate change: the development of new technologies to achieve the large-scale removal of CO_2, methane, and N_2O from the atmosphere in a safe, timely, and energy-efficient manner.

If something like this came to pass, we would have to wonder whether we were witnessing the appearance, for the first time since the invention of the steam engine, of market-competitive renewable energy. Such an event would represent the point where renewable energy begins to inhabit the same economic and technological place in human society that fossil fuels did in 1709. In other words, it would signify that the growth potential of clean energy is as vast today as was the potential of coal, oil, and gas at the start of the eighteenth century. Equally or even more important, it would also signify that the world is now poised to change in a dramatically positive way in its capacity to mitigate the effects of climate change.

From the long view of human history, I argue that if this were to come to pass, it would amount to a new energy inflection point: the moment at which we finally turn away from fire and combustion and toward energy sources vastly more suited to our long-term well-being as a species. It is also at least theoretically possible that this third moment may be inevitable. It is a frightening thought, but not an outlandish one—given our current course and the grim forecasts we face—that this moment may come to pass even if we fail to reach it deliberately. Worst-case scenarios about how climate change might occur, unrestrained by any inhibition in greenhouse gas emissions, depict reinforcement loops that rapidly escalate temperatures and initiate a multitude of other severely disruptive dynamics. Energy delivered us from the Stone Age, and we would be fools to believe it couldn't put us back there, too.

CHAPTER 10

WHAT MUST BE DONE

CORRECTING OUR CAPITALISM BEFORE IT'S TOO LATE

Solar's emerging market strength is not near enough; much more must be done to transform capitalism from driver of climate change to our best defense against it.

For many there is a pessimism—for some, a pessimism bordering on hopelessness—about the question of whether we will get our climate change act together before truly disastrous consequences are unavoidable. I see this in people I know who worry about the issue and who have a deep sense of foreboding about what they read in the paper and see for themselves in the changing world around us. In some measure, this kind of thinking stems from a very reasonable assessment of the facts today. Information about the scope of harm we know is in store for us and the persistent momentum behind the forces causing it, combined with the paltry political will apparent at present for addressing it, paint an exceedingly bleak picture. There is plenty in today's headlines to support this viewpoint, but there is also a very real path forward to a *different* future.

The political will to act is the key element, the one irreplaceable ingredient in any effective response, and on this point, we seem to have forgotten that this matter is not as entirely novel as it may seem. There are at least a few important precedents in modern times for resolving the same kind of seemingly overwhelming, entirely man-made environmental disaster we face today, although on a smaller scale. Take, for instance, the ozone hole, a catastrophe-in-waiting quietly addressed in the 1990s when the international community came together to agree to rapidly diminish the use of chlorofluo-rocarbons in enforceable ways. Or, more colorfully, consider the less well-known story of a modern civilization wallowing in its own industrialization-driven waste for centuries, in search of the political will for resolution and finally finding it. The crisis? The Great Stink.

The Great Stink is the charming name that newspapers of the day gave to the time in 1858 when the bustling metropolis of London endured both an extended drought and a heat wave of over 100 degrees, at a time when the River Thames was practically the sole sewer system for the city's approximately three million inhab-itants. More broadly, it refers not just to London's struggle to find a way to sustainably deal with its growing tidal wave of human waste but also to that of Paris and many other urban centers that emerged in the nineteenth century and had to come to terms with the consequences of industrialization and urbanization.

If you have ever been to London and seen its gleaming sky-scrapers and the sparkling River Thames—today said to be the cleanest river in the world running through a major city—it is hard to imagine the staggering crisis the city endured prior to con-struction of its phenomenal underground sewer system, one of the first of its kind when completed.

Booming as a result of the Industrial Revolution that was then surging across the country, London had grown by leaps and bounds from around one million inhabitants in 1800 to three mil-lion by 1850. The flushing toilet had become increasingly popular and was in fairly wide use in London by 1850, but all waste drained directly into the Thames, its many tributaries, or any of 200,000 cesspits dug around town.[1] When cesspits were outlawed in 1847—in a misguided effort to address the larger situation—even more sewage flowed into the Thames.[2] Ironically, widespread use of flushing toilets, which were much more hygienic than the cesspits they displaced, only compounded the problem because they ramped up overall sewage water flows, flooding cesspools and waterlogged surrounding soils and ultimately increasing outflow of sewage into waterways and the Thames.[3]

The Thames had been used in more or less this way since well before 1600—the first sewers in London, built during the Roman occupation, flowed into the river[4]—but the quantity of waste entering the Thames by the nineteenth century had long since overwhelmed the river's ability to naturally process it. The cumulative effect of the river being filled with human, animal, and other waste for hundreds of years, together with the surrounding cesspools and waste thrown onto streets, was, among a great many other unpleasant things, a fierce, omnipresent stench that became greatly magnified in hot weather.

The unpleasant smell, so powerful that just breathing it caused many to retch and gave them headaches, was only the beginning. In addition, methane and other flammable gas buildups emanating from the waste here and there around town were so numerous and intense that they periodically caused injury and loss of life—poisonings, asphyxiations, and even explosions from time to time.[5] Even worse, the concentration of waste throughout the city inevitably resulted in contamination of drinking-water sources, making them a massive petri dish of all manner of waterborne ailments, including the deadly diseases cholera and typhoid. Cesspits, in particular, provided a means of migrating contamination into drinking wells, as did the continuing practice, for many, of drawing drinking water directly from the Thames. One historian reports that, during the 1830s, "only half of the babies born in Europe lived to the age of five; the other half died of diarrhea, dysentery, typhoid, and cholera because sewage contaminated their drinking water."[6] Major cholera outbreaks in London between 1830 and 1860 claimed some forty thousand lives.[7] Contamination of the water supply was the cause, although at the time the science was poorly understood and it was popularly believed that the stench itself, or "miasma," caused the illness.

The odor, disease outbreaks, and other ill effects progressively worsened, and although piecemeal actions were taken, the will to address the crisis in a meaningful way was lacking. This was due in part to the political majority's embrace of those erroneous "miasmist" views, which tended to focus government action on addressing the smell instead of keeping the water supply clean. But the main reason for the failure to act was pure politics. London's government was organized not in a centralized manner but through an array of constituent entities called parish vestries. So grave was this political problem that the *Times* asserted that "there is no such place as London at all. [It is] rent into an infinity of divisions, districts and areas," some "three hundred different bodies deriving powers from about two hundred and fifty local Acts."[8] The self-interest of these vestries was highly localized, which created a "tragedy of the commons" problem in which there was little support for the solution required: an inconceivably expensive new sewer system for the entire city, the likes of which had never been built before.[9]

Circumstances finally conspired in 1858 to force action. The heat and drought of that summer resulted in something more than the normal stench. The Thames' water level was so low that its banks were exposed, and along with them vast areas of raw sewage slowing baking in the sweltering sun. This was more than even the most hardened Londoners could take. One newspaper wrote, "We can colonise the remotest ends of the earth; we can conquer India; we can pay the interest of the most enormous debt ever contracted; we can spread our name, and our fame, and our fructifying wealth to every part of the world; but we cannot clean the River Thames."

One wonders how much difference another development may have made to the events that followed: the original seat of Parliament, the Palace of Westminster, had burned in 1834, and in 1852 both houses of Parliament were reseated in new chambers, which were directly adjacent to the Thames. The odor inside the new buildings was so bad that one MP "rose to ask the Chief Commissioner of Works what steps he has taken, or proposes to take, to preserve the health of the members of the two Houses of Parliament from being destroyed by the present pestilential condition of the River Thames."[10]

That summer, the political logjam between the vestries and their opponents in Parliament was finally resolved, and the result was approval of massive new infrastructure to manage sewage waste by directing it out of the Thames and out of the town center in general. Including 82 miles of main sewers and 1,100 miles of street sewers, it took sixteen years to complete.[11] The master engineer in charge of the project, Joseph Bazalgette—who is today the worldwide patron saint of large public-works projects—wisely oversized his project to provide not just for current populations but future ones as well. "We're only going to do this once and there's always the unforeseen," he famously said. He was right to do so: his works continue to serve most of London today, works so vast they would cost between £50 billion and £60 billion were they built today, according to Thames Water.[12] Other major cities, such as Paris, Frankfurt, and Chicago, quickly followed the same path as London to solve their own similar problems.

It's not hard to reconsider the Great Stink as a regional version of what we face today globally with the crisis of climate change. Instead of industrialization overwhelming us with human waste, we are overwhelmed with waste from the energy we consume. Instead of treating the Thames like a free sewer for all this waste, we treat the sky as one. Instead of tens of thousands dying across the city over centuries, nearly nine million die per year from fossil fuel air pollution[13] and a multitude more are predicted to perish all around the world from the many changes coming. Instead of an array of local jurisdictions prioritizing parochial concerns above the common good, the nations of the world engage in this behavior. Instead of the political will to act arriving when lawmakers feel personally, physically at risk—well, who knows? Thus far, in our time, political will is exactly what is most lacking.

The story of the Great Stink, from its Roman beginnings to its final resolution, shows how political will finally arrived to address a crisis similar in many ways to that which we face today. If a threat to the personal safety of legislators is what will be required for serious action on climate change, we can lament that CO_2 does not reek like raw sewage; what a devilish quirk of chemistry and human physiology that its nature is to be both odorless and invisible to us. Perhaps we must await a hurricane flooding the streets of Washington, DC, or even the White House and the Capitol Building, before our leaders call upon the head of the EPA to take action that will address the threat.

■

When we are ready to finally, fully engage with the problem of climate change, what should our approach be? How can we respond most effectively?

Much of the existing conversation about how to approach climate change is simplistic to the point of being unhelpful: Drive a hybrid or electric vehicle instead of a gasoline or diesel-fueled one. Conserve energy. Support clean energy. Oppose coal plants. Eat more vegetables. Eat less meat. This conversation needs to change. While all these things are both important and helpful, talk such as this masks the true scale of the issue and the nature of the steps actually needed to mitigate and adapt to the changes that are coming.

The steps needed to truly engage require that we change the way we think about energy, climate, and economics—that we be as honest with ourselves about the many harmful ways our civilization, all across the globe, is contributing to the problem as we are brave in imagining the ways we can address it. The proper response compels all of us to answer hard questions about the way we live, how we value energy and our current climate, and how these values manifest in our economic activity. The response also requires other things, some perhaps surprising: realism, creativity, and compassion.

We must acknowledge the role capitalism has played in creating climate change, and the role it must now play in combatting it. Most people view climate change as mainly a scientific and environmental issue. The rate at which glaciers are melting, the kind of sea-level rise we can expect, the number of extinctions, the different types of species migrating to new places—these are the kinds of issues most people think of when they think of climate change. However, as important and informative as this lens is, and as useful as it has been in illuminating the situation, its serious downfall is that it focuses mainly on the trees and barely at all on the forest. As a scientific and environmental matter, the details of climate change are complex, but the main point is exceedingly straightforward: greenhouse gas emissions are causing it and the solution is to reduce or eliminate them. This technical view is very limited, however, being devoid of the crucial context the bigger picture provides. More importantly, this phenomenon keeps many from understanding properly both the extent of the problem and the ways in which we must address it.

It is time we start speaking of climate change as something much bigger than coal power plants, drowning polar bears, and shrinking glaciers. For better or worse, it is time to call it what it is: a crisis of capitalism itself, the bedrock upon which more or less all of modern human civilization rests, that is putting everything we value at grave risk.

As recounted in chapter 9, we were doing fine as a species until we opened the Pandora's box of fossil fuels and yoked them to capitalism, the economic philosophy that has largely driven the way people around the globe interact. The essence of capitalism—let the market decide the most efficient product or service and reward those providing these products or services with profits—is obviously intuitive enough to basic human nature to have been embraced everywhere in the modern world. There could be a debate about how much it has been intermingled, here and there, with socialism or communism, as well as exactly how effectively it has served some parts of humanity, but there can be no debate about the fact that, when it comes to climate change, capitalism's fingerprints are everywhere.

Since the Industrial Revolution began, energy has been the intrinsic component in almost every good and service bought and sold, everywhere. Whether you are manufacturing a widget, flipping a burger, designing a new house, or coding a new social media app, the odds are excellent that you wouldn't be doing it without electricity having played a fundamental role somewhere in the value stream of your product or service. In this way, commerce has been capitalism's handmaiden, linking the human drive to improve one's lot in life to the consumption of energy, and thereby to ever-increasing greenhouse gases. For much more than a century, we lived in blissful ignorance of this fact—although one could point to Eunice Foote's 1850s work and Svante Arrhenius's 1895 conclusion that "the advances of industry" would eventually cause CO_2 to increase global temperatures "to a noticeable degree in the course of a few centuries"—but, by the 1960s, many people knew what Exxon's findings already then showed, which was that big problems would ensue from fossil fuel emissions. Awareness has only grown since then. We have known for a long time now the likely outcome of the massive worldwide deployment of combustion-based technologies.

One of capitalism's virtues is that it is good at correcting course. If a merchant shows up in the marketplace one day selling food that sickens people, no one buys from him the following day. This is the market working efficiently: information goes to the consumer, enabling her to purchase the best good for the money. This capacity is improved by government and laws. When a shopkeeper imports cheap goods from a place rumored to have the plague, and then people get sick, the government bans imports of those goods to reduce the risk to those it governs. However, the correcting capacity is only as good as the information available, as well as the government's capacity and will to act. When both are inhibited, the machinery of capitalism grinds on, but to an inefficient outcome.

The nexus of capitalism and climate change is a giant rip in the theory of efficient markets. This rip goes to the Achilles heel of capitalism as an economic philosophy: that it requires markets to be designed properly in order to reach the proper outcome. Cigarettes provide a useful example. The cost of producing a single cigarette is likely less than a dime, but the health consequences of socialized insurance premiums and taxpayer support are a great multiple more per cigarette. The problem is that the market price for cigarettes bears no connection to their foreseeable costs, and the solution has been to vastly increase taxes on cigarettes to serve as a proxy for these costs. This is why cigarettes today cost between $6 and $14 a pack, depending on the state you are in, rather than the $2 the production costs might indicate.[14]

The same problem has happened with climate change. We have long known the likely outcome of the massive worldwide deployment of combustion-based technologies, and yet we have persistently declined, with very few notable exceptions, to correct the market by pricing combustion to reflect its likely adverse aftereffects. There are many reasons why, some of which are detailed in the excellent book *Merchants of Doubt* by Erik M. Conway and Naomi Oreskes, chronicling vast efforts to delude the masses and to slow government action—but whatever the case, it is long past the time for excuses.

At its core, climate change is much more than the very predictable result of an idiosyncratic capitalism, an errant capitalism, or a capitalism with an energy sector gone awry. It is fundamentally the logical conclusion of a capitalism that defines its markets in ways that price at zero its most prolific, predicted, and destructive byproduct: the known atmospheric consequences of its most prolific growth engines (above all electricity, but also agriculture, chemicals, transportation, and manufacturing). In the parlance of economic theory, greenhouse gas emissions have been "externalities"—for the most part, we have literally kept these consequences external to the workings of our markets. In plain English, we have allowed people to take actions the results of which harm us all, and we have not done what our markets strive to do in every other such instance, which is to ensure that those causing a harm account for it. For a long time and in many ways, this practice has done us a lot of good: we have reaped the benefits of burning fossil fuels worldwide in ways that supercharged our species' numbers and quality of life, all at a massive discount since we paid nothing for its known consequences. But the bill is coming due for our three-hundred-year tryst with fossil fuels, and it is about to do us a great deal of harm.

Our electricity markets function efficiently and with dazzling complexity in solving for the cheapest priced energy. Imagine if, instead, they solved not just for the cheapest, but also the *cleanest*, energy. If our markets were recast in this way, no one would reasonably expect them to operate any less effectively, but instead of putting off the full bill as we have been, we would be paying the bill as we go.

The uniqueness of the Barilla plant is that it demonstrates, for the first time, that solar can be both the cleanest and the cheapest energy on the market, even given the handicapping our markets provide greenhouse gas-emitting competitors. Solar's arc of declining costs and increasing efficiency shows how this trajectory is set to continue. However, the overall growth opportunity for solar and all the other clean-energy alternatives would be dramatically improved if its fossil fuel competitors were made to account for the known consequences of their emissions. This change would significantly widen and accelerate the surge of solar and other zero-emissions technologies into the market. It would also drive corrective changes throughout the economy by unmasking the market failure perpetrated by policies that gave greenhouse gas emissions a free ride.

To the extent that the ways in which we define our markets and enforce market rules are matters of public policy, the bad news is that we have no one to blame but our political leaders, who make the laws that define our markets. Of course, in democracies, it is we ourselves who chose those leaders, and therefore we ultimately have ourselves to blame—assuming no foreign power has intervened, at least. The good news, however, is that it is equally true that in democracies—and everywhere else, in one way or another— the governed have the opportunity to replace the governing every few years. It is high time that we demand that our governments play the role they should and correct the poison at the heart of our capitalism: the failure to price greenhouse gas emissions with the burden of their own consequences.

Those who are skeptical that energy markets could ever be transformed into a potent and effective tool for combating climate change may hold such an opinion because they note, entirely correctly, the role these same markets played in creating the crisis in the first place. However, this viewpoint overlooks a fundamental aspect of the shared histories of human civilization and climate change— and possibly their futures, too. If you attempted to devise a way to cause a single species on earth to engage in collective action of some kind so productive of a certain output that it affected the very chemistry of the atmosphere—all 5.5 quadrillion tons of it—you would be very hard-pressed to come up with a more effective plan than to introduce capitalism and fossil fuels to human ingenuity.

The story of climate change is essentially the story of a single phenomenon—humans harnessing combustion of fossil fuels to do work—that placed in our hands, in one iteration or another, a capability to advance, improve, escalate, and expand our capacity to execute the multitude of things we are driven to do. Not surprisingly, this technology was rapidly and pervasively adopted across almost all of human society. We are an enterprising, ambitious, and resourceful species, and the extraordinary innovation's many benefits swiftly ensured its global propagation. That the technology also created atmospheric emissions in such quantities that it altered the climate in merely a couple of centuries was incidental—and, in any case, this circumstance was largely, and willfully, ignored as we went about accomplishing the universe of fabulous new things that it made possible.

If we are looking for ways to reproduce the same phenomenon now, but for a beneficial effect—to coordinate all of humanity in achieving a second change in atmospheric chemistry—it makes all the sense in the world to look to the same thing that allowed the massive feat to occur in the first place.

Reforming our system of capitalism so that it properly prices the consequences of our emissions would create a flow-through effect, shifting all manner of activity away from the sources of emissions: consuming foodstuffs grown closer to home rather than flown in from around the world, for instance, and converting cars from gasoline and diesel to cheap renewables stored in batteries. This change would establish the means of aligning the great value of the energy we consume with the significant harm caused when emissions are released in creating it. If we reform our markets in this way, we transform humanity's deep reliance on energy from a driver of disaster into a powerful means of reducing and then eliminating the disaster. If we then extend the changes from imposing costs onto emissions to creating a price for their safe removal from the atmosphere, we would greatly catalyze the development of technologies to accelerate our progress.

In short, the union of capitalism and combustion technologies was the only thing powerful enough to create climate change.

Uncoupling this union and redirecting capitalism to the job of lowering and even removing emissions is perhaps the only thing powerful enough to remedy it. Put differently, if capitalism has caused the incredibly unlikely outcome of our species introducing a rapid change to the entire planet's chemistry, then a reformed capitalism is very likely our best hope for doing so again.

Insofar as capitalism's basic economic philosophy represents an entrainment of self-interest into our economic systems, it is now the most powerful force on the planet—basically, the drive of billions of people to do whatever they can each day to provide for their families and to achieve all their hopes and dreams. Meaningfully altering our markets in this obvious way redirects this potent force in the direction of good. Making things that create harmful emissions more expensive, and making things that reduce them more valuable, will allow markets to properly price goods and services and thus drive emissions in the right direction. Equally important in introducing this policy change is tailoring it to recognize the fundamental importance of energy to basic everyday functions in our lives and ensuring that the burden of carbon costs is not regressively imposed on those already struggling to make ends meet.

Because of all it can do, this step alone is the most powerful item to be accomplished in addressing climate change. Politically speaking, the odds of this step being taken seem long today. In fact, there is already a surprising degree of political support for the idea. A small but significant number of individuals and organizations across the political spectrum already favor putting a price on carbon emissions. Not just President Biden but also Bernie Sanders, James Baker, ExxonMobil, and the Rand Corporation, to name a few, are on record now supporting pricing carbon. Anything that could bring together a group this diverse already has a broad and politically diverse base. Many countries, including the United States, Canada, Japan, and several in the EU, among many others, already have limited policies in place to price carbon in one way or another. None are the bold and uniform global carbon pricing regime that is required, but they are significant indicia of momentum around this worthy idea. What is lacking is an upwelling of public support and perhaps a few passionate leaders to advance the cause.

We must think big to craft a muscular—and clean—energy response to climate change driven by market forces. The phrase "Energy too cheap to meter" was coined during a prior period of trial and crisis in our country. In a 1954 speech extolling the titanic effort to rapidly convert nuclear weapons technology into civilian power plants, Atomic Energy Commission chair Lewis Strauss described a bold vision for a new world in which nuclear power created so much clean, cheap energy that it was not worth making people pay for. "It is not too much to expect that our children will enjoy in their homes electrical energy too cheap to meter—will know of great periodic regional famines in the world only as matters of history—will travel effortlessly over the seas and under them and through the air with a minimum of danger and at great speeds. . . . This is the forecast for an age of peace."[15] Strauss' dream for limitless energy making possible all things we might hope for to improve our world did not come to pass, of course, but the vision remains, and in the shadow of climate change it is even more important today than it was then.

It is abundantly clear that we will need vastly more energy to manage the expansive challenges of the coming years—all of it clean. This additional energy will be required not only to maintain and increase the quality of life of the population (at its current size, and as it grows) but also to execute effective adaptation and mitigation strategies. Now is the time to pursue the means to make clean energy—renewables, not nuclear—truly "too cheap to meter," plentiful enough to meet the voracious needs we have to overcome the challenges we face.

Energy has been our most potent assistant for the past 200,000 years. This has never been truer than it is today, and we must embrace and plan for the fact that we will need much more of it as the negative impacts of climate change mount. This is why it is so important, first, to get the markets right by correcting the price of greenhouse gas-producing activities. If we succeed here, we can expect to see power plants like Barilla and other new, clean producers of energy thrive and multiply around the world in the way the steam engine once did. If we do not do this, more of the market's calls for greater energy will be met instead by fossil fuel power plants, which will cause our climate problems to spiral even further out of control.

Regarding adaptation, consider the highest-confidence element of climate change forecasts: significant temperature increases along the most densely populated latitudes. These increases, paired with anticipated population increases, are projected to drive staggering expansions of energy consumption difficult for us to even imagine today. In 2000, the world consumed about 300 terawatt-hours of energy for air conditioning, and scientists forecast this amount will grow by more than twelve times, to 4,000 terawatt-hours, by 2050, and then incredibly by more than thirty times, to 10,000 terawatt-hours, by 2100.[16] The need for relief from warmer and warmer temperatures is only part of the equation; rising incomes in places like India and China, lifting millions out of poverty, is another driver for increasing air conditioning demand.

Adaptation strategies to deal with water supply and infrastructure will also drive greater energy consumption. Fresh water scarcity is forecast to require a massive expansion of ocean desalination infrastructure. The largest cost of desalination, by far, is its power bill, due to the vast energy requirements of this technology. Coastal inundation will likely require energy for increased pumping activity to stave off rising waters. Consider, for instance, that even today, New York City's subway system pumps thirteen million gallons per day just to keep the tracks clear.[17] As ocean levels rise in and around subway lines and other infrastructure in New York and other coastal cities around the world, invading waters will drive larger and larger electricity consumption just to keep the systems functioning.

Technology to make possible perhaps the most important mitigation strategy of all—the cheap and safe removal from the air of vast quantities of the emissions previously released—has not been invented yet, but there is reason to believe it will be energy-intensive if and when it is.[18] Almost all future scenarios examined by the UN IPCC in which temperatures are stabilized require that we not only reduce emissions to zero at some point in the next few decades but also that we "go negative," i.e., begin removing CO_2 from the atmosphere at scale. Technologies to accomplish this on the order of magnitude required do not offer much promise at present; designing them will probably only be possible through a historic research and development effort, something on the scale of another new Manhattan Project.[19] What is known from work done to date, however, is that the "direct air capture" technique, among the more theoretically scalable methods, requires significant energy expenditure due to the unique molecular properties of CO_2, which make it difficult for it to bond to other elements. Put more plainly, the only ways we have come up with to date to remove carbon from the air at scale are incredibly energy intensive, and if we can't find a way to meet these energy requirements from something other than coal and natural gas power plants, we will never succeed.

The energy required to power the adaptation and mitigation strategies I list here is not included in current forecasts of future energy demand, but in my view they are the ones that must be planned for with substantial urgency. They will all be crucial for maintaining life and property as populations increase and climate change ramps up, and all are likely to come with massive energy consumption footprints.

These new demands for power mean that we must think big, and creatively, about how to get the energy we need in the future. What does thinking big and creatively look like? Barilla has shown the way to vastly increase cheap energy supplies by going solar, but solar must be reimagined as the core of a new approach to energy, in which it is paired with other cheap renewables in the context of weaning the world off the vast fossil fuel infrastructure it relies upon today. In time, deployment of proven zero-emissions technologies like solar and wind and of new ones not yet proven—at the terawatt, not megawatt, scale—into resource-rich sun and wind areas around the globe (such as the western and central United States, the plains of Canada, Russia, Mongolia, Western China, Australia, the Sahara, and the oceans), linked with plentiful new transmission infrastructure and promising new energy storage technologies like batteries, liquified air technologies, gravity systems and hydrogen derived from renewable energy, will be eminently feasible, should the political will materialize. In addition, large R&D investments should be made in technologies that exploit the extraordinarily energetic and 100-percent-reliable resource covering about 70 percent of the planet—the tidal and wave energy found in the oceans—which, to date, have largely resisted innovation.[20]

Those who care deeply about climate issues may be tempted to support a simple command-and-control approach, with the government zapping coal and natural gas power plants while ordering new solar and wind projects to be built here or there. But this would be a terrible mistake, given the vast complexities of grid systems, the many factors that drive diligent infrastructure execution and the incredible accelerant private capital would be to the achievement of our climate imperatives. The best path, instead, is to fundamentally reform our energy policies and markets without delay so they drive investment into clean power generation and out of assets that create damaging emissions so they may find their place, if there is one, in a world engaged in a serious effort to manage climate change.

Putting a price on carbon emissions is the key to achieving rapid energy-sector decarbonization. Once emissions are priced properly, power plants that create them will either survive or not; if new technologies arise to make their operation economic, such as carbon capture, so be it, and if not, they will exit the space in an orderly way for reasons related to their own finances. New, cleaner power plants will be drawn into the market to fill the gap, under procedures that meet the complex needs of grid operators charged with ensuring that supply equals demand for every consumer across their service area. Although this kind of process will unfortunately require time to produce the categoric changes we need—longer than any ideal schedule for managing emissions would allow, to be sure—I believe it will hasten the path to securing massive new investment in building cleaner generation infrastructure better than any other. To successfully put capitalism to work for the climate's sake means observing the rules and norms that make capitalism effective, such as transparency, predictability, and active and impartial regulation backed by the rule of law.

One need only look to the example of China and its command-and-control approach to renewable energy deployment over the last decade to see the hazards of having bureaucrats determine renewables capital investments. China's centrally planned decision-making has resulted in large quantities of wind turbines being constructed but relatively small amounts of wind power being delivered.[21] This odd outcome arose from government policies that supported projects being built in areas without sufficient transmission access or adequate wind speeds, and with wind turbines not suited to the conditions of the sites. The result was a large amount of precious capital being deployed for projects that deliver about 25 percent less than the energy that wind projects produce in countries with market-driven economies.[22]

This approach would likely mean that natural gas power plants will continue to be a substantial part of our energy mix for some time, given the low price of natural gas and its increasing deliverability to places around the world otherwise looking to coal. Many climate advocates struggle with the issue of what role natural gas should play in a responsible energy plan. On the one hand, all energy generation producing greenhouse-gas emissions should disappear as soon as possible, and the natural gas extraction and distribution processes are drivers of fugitive methane emissions disastrously accelerating greenhouse gas accumulations; on the other hand, natural gas is much cleaner than coal in terms of emissions, is cheap and fairly plentiful, and is a power source that can be turned on and off when needed so that it makes an agreeable fit with renewables that come and go with the sun or wind.

My hope is that fugitive methane emissions are addressed immediately so that natural gas generation can remain a resource available when we need it to keep the lights on in homes, schools, offices, and hospitals as the overall proportion of renewables grows higher and higher, a role I fully expect to fade away. If markets are corrected, lower- and zero-emissions power sources will rush in even more quickly than they are doing today, and natural gas plants will become less competitive as they shoulder the burden of their CO_2 emissions. This process has already begun even without making natural gas plants pay for their emissions, but pricing carbon will dramatically accelerate it. Energy storage technologies are rapidly evolving to serve these same purposes in making renewables better fit our demand for constant power supply and soon will compete successfully with gas plants to fill the gaps that renewables leave. This transition would happen over a period that would allow planning, investment, and construction of clean-energy assets in a way that is orderly enough to maintain grid reliability. Eventually, if storage technologies follow in the footsteps of solar in becoming cheaper year after year, we will see in as little as a decade gas plants become mainly emergency backup resources.

We must be aggressive, but also realistic and pragmatic, about how we shift from fossil fuel–dominated to renewable energy–dominated generation. Between the complexity of scientific issues and the vitriol of politics, the importance of moderating principles in mapping our response to climate change is easily lost. Many decisions will need to be made in the gap, with incomplete information, and these principles will play an outsize role in guiding the matter of America's response to the changes coming. Pragmatism will be key. Any realistic view of the problem of energy sector emissions has to acknowledge that billions of people worldwide depend on electricity and transportation from fossil fuel combustion for life and livelihood, and that the vast majority of the world's existing energy infrastructure is simply not going to change instantly. Cumulative investment in fossil fuel–generating facilities around the world totals in the trillions of dollars, with its value reflected in publicly traded stocks held by pensions, mutual funds, and private individuals, and no venture this large turns on a dime.

US Senator of West Virginia Joe Manchin, a Democrat who believes climate change is real but who represents a large coal state, articulated this point very well when asked in 2016 about candidate Donald Trump's rejection of climate science. After calling out Trump and others like him as "deniers," he went a step further and also called out "deniers" on the other side. "[W]e have deniers that believe we're going to run this country or run this world without fossil. That's a worse denier, thinking they're just going to just shift it and everything's going to be hunky-dory. . . . Can you imagine what would happen if you start to have rolling blackouts . . . because the grid is not charged and there's not enough energy in the grid system?"[23] As dire as the climate situation is, Senator Manchin was not wrong to call out unrealistic scenarios in which our emissions problems are quickly solved without diminishing the reliability of our power infrastructure and putting the health and security of people everywhere alive today at risk. So many of us, particularly Americans, take reliable power for granted, but as an increasing number of events prove to us each year, losing it quickly becomes a matter of life and death. The polar vortex event in Texas over five days in February 2021, for instance, resulted in at least 246 deaths, many from hypothermia and others from asphyxiation due to CO_2 inhalation.

Nuclear power is another conundrum for some climate advocates. While nuclear power plants can produce vast amounts of zero-emissions energy, despite Lewis Strauss' bold dream for cheap power nuclear is just not competitive with natural gas, wind, or solar generation in today's electricity markets. The nation's aging nuclear plants—all but three of the sixty-one operating plants in the United States were built at least twenty-five years ago—not only find themselves losing money more often but also carry the massive financial liability of having to dispose of their nuclear waste, not to mention one day decommission and remediate their operations sites, two immense burdens that are only partially covered by federal insurance. Together with the onslaught of cheap renewables, this situation has resulted in a growing wave of nuclear plant closures, including eleven since 2013 and eleven more announced for the coming years.[24]

Existing nuclear plants are suffering from competition, and the new plants being built today are certain to suffer the same fate. During the Obama administration, a major effort was made in the financial crisis stimulus response to revitalize the nuclear industry. Funds and loan guarantees were provided to build new nuclear plants in Tennessee, Georgia, and South Carolina. If the experiences of these plants being constructed are any indication, the future prospects of nuclear energy are extremely daunting. Only one of these plants has been completed on schedule, the Watts Bar 2 plant in Tennessee. This plant saw its original cost estimate of $2.2 billion rise to $4.7 billion by the time it was completed,[25] and after opening it was more often shut down for safety reasons than it operated.[26] Of the remaining new US plants, Vogtle 3 and 4, the twin Georgia plants, were originally estimated to cost $14 billion, but overruns have pushed them, thus far, to over $25 billion,[27] and V.C. Summer 2 and 3, the twin South Carolina plants, were cancelled after numerous massive overruns.[28] These types of overruns for new nuclear plants, a terrible problem for the utilities building the plants and more importantly for their ratepayers, are only the beginning. The overruns are so vast, they have raised concerns about the viability of the entire US nuclear industry. Westinghouse, a unit of Japanese conglomerate Toshiba, was the contractor for each US plant. The financial burdens of the overruns caused Westinghouse to file for bankruptcy protection, and Toshiba, which provided guarantees for its subsidiary's work, was nearly drawn into bankruptcy as well.[29]

Wind-, solar-, and gas-generated electricity can all be purchased for a fraction of the price of nuclear, and nuclear's pricing problems only look to be getting worse; meanwhile, its advantage as a baseload generator is being overcome with rapidly advancing clean storage technologies and transmission innovations. If you like clean energy, you can get it much more cheaply from a solar or wind plant than from a nuclear plant, and if you don't care if it's clean, you can get it more cheaply from a natural gas plant.

We must be forthright about what we are facing and begin speaking of climate change as our definite future, not a theoretical one. The increasing frequency of natural disasters, droughts, and heat waves is driving greater awareness of climate change, but it can still be difficult for many to get a view of what these changes will mean for our future. One reason that gaining this understanding can be hard is the widespread phenomenon of compartmentalizing climate forecasts and separating them from other projections about the future. Climate change forecasts can seem to exist in a distinct universe from business-as-usual predictions, as though the set of changes that virtually all scientific inquiry on the matter tells us we can expect is hypothetical. By projecting separate truths to choose from, we make climate change projections seem like alternatives, instead of a given.

Consider the matter of the US government's economic growth forecasts. You would think, given the number of climate reports the federal government has published over the past twenty years, that our economic forecasts would integrate all of these forecasts into its economic predictions. After all, what would be the point of producing all those reports if no one uses them? In fact, however, our government's growth forecasts take almost no account of climate change, even though our economic projections go out as far as 2060.

The US Office of Management and Budget (OMB), the federal agency with primary responsibility for executive branch budget-making, annually forecasts economic growth based on its extensive modeling, forecasting, and economic trends analysis. Each year these findings are summarized in the very bureaucratic-sounding *Analytical Perspectives* publication, a voluminous document putting forward economic growth forecasts together with an analysis of the president's proposed budget for that year. The 292 pages of President Trump's first *Analytical Perspectives* document mention climate change only once, when discussing risks to farming insurance programs,[30] and do not refer to climate change at all in the key chapter detailing economic assumptions and economic growth forecasts for the years 2017 to 2027. One presumes the Biden administration will do better, but it is also true that the Obama era reports were not particularly robust in this respect.

Congress isn't much better. The Congressional Budget Office (CBO), the nonpartisan office responsible for Congress's economic forecasting, takes the same approach as Trump's OMB. The CBO's main annual forecast, "The Budget and Economic Outlook: 2017–2027," does not mention the words "climate change" even once in its 130 pages.[31]

One could question whether the quantum of change expected to occur over just the next ten years will be of a magnitude to have any effect at all on economic growth. In fact, this very question was considered in a 2015 survey of economists about the effects of climate change on economic forecasting. The survey found that more than 40 percent felt that climate change is already negatively affecting the economy, and the median estimate was that it would be having a negative impact on the global economy by 2025. These economists predicted that major sectors of the US economy will be harmed by climate change, including agriculture, fishing, utilities, forestry, and health insurance—all major US economic players. The survey also found that large majorities of the economists surveyed felt that many modeling practices and norms tend to understate the impacts of climate change in economic forecasts.[32]

The practice of segregating the consequences of climate change from long-term economic projections is not unique to the United States government. The Organisation for Economic Co-operation and Development (OECD), an intergovernmental body of developed nations, has authored many reports about the impacts of climate change, and many forecasts of economic growth as well. One of its most innovative efforts was a report entitled "The Economic Consequences of Climate Change," published in 2015 in the run-up to the UN conference held to inform parties negotiating the Paris Agreement. Alarmingly, this study concludes that "in the absence of further action to tackle climate change, the combined negative effect on global annual GDP could be between 1.0% [and] 3.3% by 2060 . . . [and] between 2% and 10% by the end of the century relative to the no-damage baseline scenario."[33]

With such an emphatic viewpoint on the impacts of climate change on economic growth, one would certainly expect anything else published by the OECD to include these kinds of findings—but this is not the case. It may be understandable that a prior comprehensive OECD long-term growth assessment report,[34] issued three years earlier, excludes climate change impacts *entirely* from its economic projections. How would the authors of a 2012 report have any idea of the large effect climate change would be estimated to have only a few years later? It turns out they would have had a very good idea had they read another OECD report published a few months earlier, in 2012, titled "OECD Environmental Outlook to 2050: The Consequences of Inaction"—the precursor to the

2015 "The Economic Consequences of Climate Change" report. This 2012 report found that "if the costs of taking no action to address the key environmental challenges were fully included in this assessment, future GDP would be lower than projected" under the baseline assumptions, which were those recited in the 2060 report.[35]

Economic forecasts are not the only ones segregated from climate change projections. Consider the United Nations' recent projections for world population on the one hand and its climate change predictions on the other. The UN's most recent world population report, *World Population Prospects, 2019 Revision*, predicts a climb from the 7.7 billion people on earth in 2019 to 8.5 billion by 2030, 9.7 billion by 2050, and 10.9 billion by 2100.[36] One of the countries projected to experience particularly remarkable and rapid population growth is Nigeria, currently the seventh largest population at about 200 million. The UN forecasts that Nigeria will, by 2050, surpass the United States to become the planet's third most populous nation with 401 million people (compared to 379 million in the US). By 2100, Nigeria's population is projected to reach 733 million, just about double the projected US population.[37]

When you consider the *Prospects* findings alongside the work of its sister UN entity, the IPCC, you wonder whether the former's authors were familiar at all with the small ocean of IPCC reports published over the years forecasting the regional impacts of climate change. *Prospect*'s forecast of a phenomenal 50 percent increase in global human population is put forth within a report that mentions the words "climate change" only four times, and only then in relation to the challenging impacts on the world's five island nations. The report does not reference many other important factors noted in the IPCC reports detailing the consequences of climate change on world populations.[38] Neither does it mention that its list of the nine fastest-growing countries through 2100 includes four developing countries—Nigeria, Democratic

Republic of the Congo, Pakistan, and Ethiopia—which are each identified in the IPCC reports as uniquely vulnerable to climate change.

It seems that the *Prospects* authors must be unfamiliar with the IPCC's special report detailing challenges that will be faced by African nations, published five years earlier, in 2014.[39] This report paints a vivid picture that is greatly at odds with the kinds of conditions suited to sustained population growth. Consider the following forecasts the report makes regarding some of the areas on the African continent predicted for strong population growth: higher temperatures and less precipitation; deterioration in food security as a result of diminished agricultural and fisheries yields; stress on water availability; and widespread amplification of health threats from diseases and malnutrition. "Climate change and climate variability have the potential to exacerbate or multiply existing threats to human security including food, health, and economic insecurity, all being of particular concern for Africa (medium confidence)," advises the report. "Many of these threats are known drivers of conflict (high confidence)." The IPCC's special report on Asia, including Pakistan in the South Asia region, contains similarly concerning forecasts of food insecurity, intense heat waves, diminishing rainfall patterns and regional conflicts.[40]

Where does it all lead? Consider how the IPCC Africa report sums up:

> Of nine climate-related key regional risks identified for Africa, eight pose medium or higher risk even with highly adapted systems, while only one key risk assessed can be potentially reduced with high adaptation to below a medium risk level, for the end of the 21st century under 2°C global mean temperature increase above preindustrial levels (medium confidence). Key regional risks relating to shifts in biome distribution, loss of coral reefs, reduced crop productivity, adverse effects on livestock, vector- and water-borne diseases, undernutrition, and migration are assessed as either medium or high for the present under current adaptation, reflecting Africa's existing adaptation deficit. The

assessment of significant residual impacts in a 2°C world at the end of the 21st century suggests that, even under high levels of adaptation, there could be very high levels of risk for Africa. At a global mean temperature increase of 4°C, risks for Africa's food security (see key risks on livestock and crop production) are assessed as very high, with limited potential for risk reduction through adaptation.

Even in the dry vernacular of the IPCC, the description of the cumulative effect of these changes on the people of Africa through the remainder of this century is chilling.

What's more, these same nations projected by the UN to be among the fastest-growing—Nigeria, Democratic Republic of the Congo, Pakistan, and Ethiopia—are all flagged in the Fund for Peace's *2020 Fragile State Index*. "Fragile State" is the term for a state experiencing pressures that "are outweighing [its] capacity to manage those pressures," with the *Index* being an annual ranking of the world's countries most at risk. Congo is number five, Nigeria fourteen, Ethiopia twenty-one, and Pakistan twenty-five out of the 178 countries assessed.[41] For reference, conflict-torn Syria is fourth, Afghanistan is ninth, and Iraq is seventeenth while the United States is one hundred forty-ninth and Finland is the least fragile nation.

Is it reasonable to expect Nigeria to sustain almost a fourfold population increase, from 200 million today to 733 million in 2100, during the same window in which the stresses of climate change will likely come to bear, as detailed in the IPCC reports, and when its government is already in an exceptionally fragile state? And are the forecasts for Ethiopia, Pakistan, and the Democratic Republic of the Congo, which are each projected to join the ten most populous nations by 2100, also sustainable? It is easier to believe simply that the left hand does not know what the right hand is doing when it comes to climate change forecasts, at the United Nations and many other places.

We can, and should, demand much better from the United Nations, one of the most authoritative resources we have when it comes to the impacts of climate change. We can, and should, also expect much more from the United States government. Until we are willing to discuss climate change as our definite future, instead of just one possible future among many, we will fail to engage honestly with the problem.

We must shun the every-man-for-himself approach in favor of a coordinated global response. In his 2016 book *Fossil Capital*, Andreas Malm puts forward an economic theory of climate change through history, with particular emphasis on the Industrial Revolution and the role China has come to play. In Malm's view, China is a modern bookend to England, in its historical role as the nation that launched the industrialization that began climate change. England may have started it, his argument goes, but China seems to be bringing it to its logical environmental conclusion.

Malm makes a fascinating point toward the end of his book that vividly crystallizes the intractably global nature of the climate change problem. He first notes that, of all greenhouse gas emissions between 1751 and 2010, half occurred in the twenty-five years between 1986 and 2010, and emissions since 2010 have continued to grow at about 3 percent per year. This historical accumulation paired with new emissions echoes the alarming picture painted by carbon budget analysis; Malm's own view is that this trajectory places us beyond even worst-case UN IPCC scenarios, setting a course for a shocking temperature increase of 4 degrees Celsius as early as 2060.

Looking for what specifically accounts for the sharp burst since the 1980s, Malm connects the bulk of the increase to China, in particular to the incredible growth in its exports following its gradual pivot to capitalism, which really took off in the 1990s. He points out that, between 2000 and 2006, 55 percent of all growth in CO_2 emissions came from China, with 66 percent in 2007, and he traces the majority of these emissions to China's exploding export sector, which was shipping ever more goods to developed nations.[42] Imports from China to the United States increased by

250 percent from 1997 to 2007, and imports to the European Union increased by 154 percent.

Malm's argument is clear: the massive burst of emissions in recent decades arose mainly from China positioning itself as the producer of goods for the rest of the world. The "combustible mix of China and globalization has set off the emissions explosion," he argues. Connecting the dots back through history, Malm writes, "If Manchester was the 'chimney of the world' in the 1840s," during the height of the Industrial Revolution in England, then "the People's Republic of China assumed that position in the early twenty-first century."

Incidentally, over this same time period, while exports were causing China's total emissions to skyrocket, the United States and Europe were heralding the fact that their emissions rates were leveling off, and even occasionally going down.[43] This suggests that, while we in the United States and Europe thought we were making pivotal improvements in our emissions, we were ignoring another part of the story—that, through global trade over a few decades, the world has simply seen the movement of many smoke-stacks from locations in the West to China. This all raises the question: Has even the modest success we thought we had made in reducing our emissions been illusory?

Malm vividly expresses the "whack-a-mole" aspect of the problem of greenhouse gas emissions, laying bare how deeply linked countries are in creating the problem we are trying to solve. The same is also true about the links that bind us when it comes to finding solutions, which brings to the fore the forces of nationalism that have emerged around the world in recent years, including, most notably, former President Trump's "America First" policies, which would have turned the United States away from the idea that climate change can and should be faced in a coordinated way.

Instead of a "we're all in this together" approach, former President Trump and those of like mind would have the world engage in an "every man for himself" response, a path representing a grave threat to effective steps to mitigate and adapt to climate change. If his approach were to carry the day, his years in office would represent the beginning of the end of a unified response to climate change, and no one should expect it was vanquished in the 2020 election. It is important to note that even before the 2016 election, consequences of climate change already manifested as centrifugal, atomizing events that have not only complicated cooperation but fomented conflict among nations and peoples. Resource disputes and large-scale refugee flows have strained international norms, for instance. In even the best of circumstances, preserving international comity would be difficult as increasingly severe climate change effects manifest. The United States, the world's only superpower, will have much to say about the course that is taken. If we go it alone, undoubtedly many other nations will see no reason not to do so as well, whereas if we push for cooperation, others are more likely to do so too.

The case for working with all the other nations of the world to collaborate on technologies and emissions reduction solutions is as obvious as it is overwhelming, and it is an approach the Biden administration has wholeheartedly embraced. There is no way in which we do not benefit by enlisting the best and brightest among all nations to collaborate and find solutions. For any threat to substantial numbers of people across the globe, nothing else makes any sense at all.

This turn of events brings to my mind an experience from my youth back in Galveston. When I was young, I played basketball for my local YMCA team for many years. A couple of those years, our team was good enough to qualify to compete in the state championship tournament in Austin. I have vivid memories of those years, because getting to go to the tournament was such a big deal—we got to leave school early, travel together to Austin, and stay overnight in a hotel. The games did not go well for our team, however; the best we ever did in the tournament was to survive for a game or two. This meant we had the opportunity to watch the better teams through the finals, and there was one team that seemed to be perennially in the running to win the whole tournament. They were called the San Antonio All-Stars, and they had gorgeous red, white, and blue uniforms with lots of stars. They looked like real basketball players and were bigger, stronger, and just all-around better than us and most of the other teams present.

Why were they so much better? My team from Galveston was just the team from our league that qualified for the tournament, whereas, as my coach explained to us at the time, the San Antonio All-Stars was a different kind of team: it literally consisted of all-stars from many teams who were combined into a new team—a collection of the best players from all across town. I'm not sure why the tournament rules allowed teams to be organized in such different ways—a team made up of all-stars was always going to prevail against the rest of us—but their dominance made an impression on me. It was conceivable that we could beat the best team in San Antonio, but we had no chance to beat the best players collected from all the best teams across the whole city.

The story of photovoltaic solar's incredible rise to market competitiveness is one in which the world's best scientists, businesspeople, bankers, and policymakers worked together to create a globally dominant product. An international "all-star" combination of critical ingredients—American ingenuity; German and Japanese technology, expertise, and policy; and Chinese manufacturing wherewithal and unprecedented financing, together with contributions from a multitude of other countries along the way—made solar what it is today. The United States could not have achieved this outcome alone, nor could Germany, Japan, or China. The rise of solar to dominance is, more than anything, a product of the world's daring experiment with globalism. Much has been written about the negative effects of globalism—economic dislocation, rising inequality, environmental degradation among poorer countries, and more—but there are also many good consequences of global cooperation, and solar is one of them.

When I heard of Donald Trump's denials of climate change and its human causes, and then of his decision to exit the Paris Agreement in 2017, I couldn't help but think about the San Antonio All-Stars and their dominance at the state tournament. If Trump's real desire was to turn his back on globalism and retreat into an isolationist posture in which we go it alone on climate change and many other issues, then we will be choosing to forfeit the advantages of an all-star team just as we head to the big tournament. Trump's desire to turn the United States inward, to alienate the best and brightest of the world who are seeking to collaborate with us in addressing the world's most challenging problems, is the biggest single threat his presidency represented. We must hope that the Trump years will be but a glancing blow to the momentum of the international community in coming together under the Biden administration and beyond in our response to climate change. But only time will tell. The next election is never too far away, and Trumpism will all but most certainly raise its head again.

Americans must recognize and address the obscure constitutional peculiarity with the massive carbon footprint: The Electoral College. On its face, it may seem quite counterintuitive that a relatively obscure part of our Constitution can be identified as a culprit in the long and complex story of climate change. Most reasonable people, when asked whether the unusual way that the United States selects its president has anything to do with climate change, would dismiss the question out of hand. Oddly enough, however, it turns out that the Electoral College has been a very significant impediment to serious climate action in the United States over the last two decades, and unless action is taken, it may continue to be over many more to come.

Consider the mechanics of Donald Trump's victory in the 2016 election, and Joe Biden's in 2020. Trump's election was not decided by the three-million-vote margin by which Americans nationwide preferred Hillary Clinton to him but by the small margin he garnered in the Electoral College. History books one hundred years from now will likely note the irony that it was just 77,759 votes for Donald Trump across America's Rust Belt—much of it coal country, historically[44]—that overwhelmed Hillary Clinton's popular vote advantage, which came in large part from "blue" states that support climate action. But for these 77,759 votes, we would not have the very significant pause, and even reversal, on climate progress that Trump's policies have initiated. Instead, we would have seen Hillary Clinton's administration continue more or less the progressive energy and environmental policies of the Obama administration, ensuring such things as our continuing participation in the Paris Agreement, a robust defense of carbon emissions regulations and vetoes of ill-considered climate legislation out of the Republican-led Congress.

The 2020 result was strikingly similar to that of 2016, but with Biden eking out the Electoral College win and expanding the popular vote advantage. In Biden's case, his tight advantage across a handful of states was all that stood between him and a loss similar to Hillary Clinton's, despite having racked up more than 7 million more votes than Donald Trump. The same metric showing Trump securing his 2016 Electoral College victory by just 78,000 votes shows Biden's winning margin in the three states that gave him the Electoral College—Arizona, Georgia, and Wisconsin—was even narrower at just 45,000 votes.[45]

In Biden's case, things worked out for the climate change candidate, but this is an anomaly. The Electoral College didn't just cost us Hillary Clinton's presidency, it also cost us Al Gore's. George W. Bush's victory over Al Gore sixteen years prior to Clinton's loss was consummated via the "hanging chad" squeaker election in Florida that gave Bush Florida's Electoral College votes, and with it victory. Those chads were enough to overcome Gore's half a million popular votes advantage.

From a climate change point of view, it is downright painful to imagine the "what if" scenario of a Gore presidency. At a minimum, we would have had four or eight years of strong and thoughtful support for climate action; in the best case, we would have seen the stirrings of a serious and coordinated global response to climate change that laid the foundation for a whole different world than the one we have now. This shift might have occurred at a time when a significant bend in the trajectory of US and global emissions was within grasp, which could have yielded extraordinary improvement over the urgent moment in which we stand now. On domestic policy, Gore was a pioneering thinker on clean-energy policies, the carbon tax, and a wide range of greenhouse gas emissions reductions approaches. Internationally, Gore supported the Kyoto Accord which, had he found a way to steward the agreement through Senate approval, would have set the world's leading carbon emitters on a path to coordinated action to reduce emissions. Instead, the treaty was dropped by the Bush Administration, setting the bold precedent for US abandonment of multilateral climate action that was recently followed by the Trump administration.

Far and away the most tantalizing what-if of all the would-be Gore presidency, however, lay in the possibility that he may have achieved the elusive factor most critically lacking today on climate policy: bipartisan support for serious action. A number of Republicans during the late 1990s and early 2000s, including party leaders such as John McCain, Arnold Schwarzenegger, and Richard Lugar, were actually in support of robust measures to address climate change in that period. Republican President George H. W. Bush took bipartisan action to address the ozone hole and signed the UN Compact on Climate Change in 1992, and several Republicans joined most of their Democratic colleagues in the US Senate in support of 2003's Climate Stewardship Act, co-sponsored by Senators McCain and Joe Lieberman.[46] This all ended during the George W. Bush Administration, which began to vilify climate action positions, setting the stage for the party-wide opposition on climate change which in time has curdled among many Republicans into scorched earth countermeasures and outright climate denialism.

In retrospect, it is clear now that this could-have-been biparti-san convergence on climate matters was in fact a brief and unique opportunity in US history—a moment entirely lost to us as a result of Gore's defeat, even though a majority of Americans cast their votes for him.

One could argue that the Electoral College should not be blamed for any part of our climate woes, unless either Gore's or Clinton's loss was mainly determined by the candidates' climate positions. If the election did not turn on the issue of climate change in either instance, the case may be made, the fact that the Electoral College turned the outcome one way or another is neither here nor there. That climate matters were not decisive issues in either election is demonstrably true; neither Clinton nor Gore made cli-mate action the centerpiece of their campaign, and in fact the issue was fairly far down any list of key election issues. Notwithstanding this fact, however, there is no denying that a majority of voters in each election supported the candidate backing action on cli-mate issues—and more importantly, had that candidate prevailed, neither can it be denied that meaningful climate progress would likely have been made.

If it seems surprising that the will of a majority of American voters was subverted in these presidential elections by the Electoral College, it shouldn't. In fact, the Electoral College was created to ensure this very possibility: that a path exists for a minority of voters to prevail over the majority in choosing our president. The system was originally conceived during the Constitutional Convention of 1787 as a way to address the fact that many of the participants desired to insulate the presidency to a degree from democracy. While there is debate about the exact provenance of the rationale for the mechanism, it is hard to overlook the thinking of James Madison, the "Father of the Constitution" who was closely involved in conceiving and drafting the language establishing the Electoral College. The Convention was challenged in balancing the diverse interests of all the states of the new nation and enlisting them to come under the nascent federal government, and a central difficulty was assuring the Southern states that they could protect their interests in electing the president of the new nation even though they had many fewer voters (since enslaved people were not allowed to vote) than the Northern states. Madison's solution to this dilemma was the Electoral College, for which he explained the need as follows:

> There was one difficulty however of a serious nature attending an immediate choice by the people [in electing the president]. The right of suffrage was much more diffusive in the Northern than the Southern States; and the latter could have no influence in the election on the score of the Negroes. The substitution of electors obviated this difficulty and seemed on the whole liable to the fewest objections.[47]

Some historians assert that Madison's reference to the lack of "diffusive" suffrage in the South was incidental and that in fact he and others at the Convention supported the system not just to accommodate the interests of the slave states but also to prevent the "tyranny of the majority" in selecting the president.[48] There is evidence to support this point of view, but it is noteworthy that the two interpretations are not mutually exclusive if the concern was that a true majority of voters then would have threatened the interests of the slave states. And, in any case, Madison's own explanation seems specific enough to be difficult to argue with.

The Founding Fathers had their reasons for stitching the country together politically by way of the Electoral College, and it has not always served us poorly. Abraham Lincoln, for instance, gained the presidency through the Electoral College with only 40 percent of the popular vote. But it cannot be denied that there have been real and negative consequences from its apportionment of outsized input for less populated states in presidential elections. The complete list of pros and cons of the Electoral College is a topic for an entirely different book, but any serious look at the history of climate policy in the United States, and its impacts around the world, would be lacking if it did not note its tremendous and negative impact. Our method of choosing presidents has unquestionably been an epically poor fit with regard to climate change. Until our presidents are elected in a way that ensures the support of the majority of Americans, we perpetuate the significant possibility of candidates ascending to this highest and most powerful office who are out of sync with what a significant majority of Americans already favor on this matter of ever-greater importance: ensuring serious action on climate policy.

That the same constitutional accommodation forged to sustain the epic tragedy of slavery, our country's first and greatest failure to follow through on our fundamental principles, has continued to bear poisonous fruit in the form of our current failure to address climate change, is inexcusable. If, as E. B. White wrote, "Democracy is the recurrent suspicion that more than half of the people are right more than half of the time," it is time that we Americans honor this principle in our presidential elections. The time for constitutional complicity in perpetuating a path for minority rule when it comes to choosing our most powerful executive has surely come to an end. As recent history has made abundantly clear, the Electoral College must be reformed, if not abolished, if we are serious about responding to climate change and much else. Similar arguments could be made to reform the US Senate, which also confines the role of national majorities by apportioning the same representation to states with vastly different populations.

■

Like parents everywhere, my wife and I wonder and worry about what the future holds for our three children. The billions of parents living in developing countries share the same worries but face much greater risks. It is a sad fact that unless we take extraordinary action, the baseline expectations for all of the lives of these children, and especially for the lives of their children and grandchildren, will be worsened significantly by the array of negative consequences already being driven by climate change. The lives these individuals have grown up expecting will differ in important ways from the ones they will experience. In even a best-case scenario, unless we rapidly implement the dramatic changes necessary to alter our trajectory, the next several generations will pass through significant disruption and deterioration of conditions before any stabilization may occur; in the many other scenarios in which we fail to make adequate changes, the disruption and deterioration will start sooner, continue longer, and cause much more mayhem.

In some ways, each generation over the last century has struggled with a version of this problem. My grandparents came of age during the Great Depression and worried that their children would face a world stunted forever by poverty and by intractable, worldwide economic stagnation—and then they faced the rise of fascism, world war, and the very real threat of defeat and the end of Western democracies. Their children, the Baby Boomers, grew up during the Cold War, hiding under desks during nuclear bomb drills and worrying that their families might be incinerated at some random time chosen by the Soviets or a Dr. Strangelove. All these respective fears were legitimately and deeply felt, as our ancestors before them no doubt periodically faced existential threats to their countries, their ways of life, and the whole world as they knew it. All their efforts to prevail over these threats were truly herculean and reshaped the world we live in today.

And herein lies the biggest distinction between these great challenges in our past and the present predicament: those prior generations confronted the dire threats facing them and eventually prevailed over them, whereas our generation, at least thus far, has not yet even seriously engaged with the problem.

CHAPTER 11

A LOVE LETTER TO AMERICA ABOUT CAPITALISM AND CLIMATE CHANGE

Major climate progress has never seemed more remote in the United States, but there are compelling reasons to believe meaningful engagement to address climate change is right around the corner.

f the story of solar energy has been a linear chronology, starting with Edmond Becquerel's accidental discovery in 1839 and continuing through innumerable iterative improvements up to today, the same cannot be said of natural events. Nature's beat through history is most often a circle or ellipse, an orbiting of events returning to the same places again and again. Among the people who likely know this lesson best are those living on islands next to tropical ocean waters—and among them, the oft-battered Galvestonians. In its recorded history, the island city has been struck by hurricanes fifty-five times in the past 147 years, and each June brings a new chance with the start of hurricane season.[1] Just as sure as each strike threatens life and property, so too will the question afterwards be posed by the survivors: How can we stop the next one? The grim reality that climate change will augment tropical storms—warmer oceans means stronger, wetter, and bigger storms—intensifies the dread these days, and the resolve to prevent future damage.

After Hurricane Ike ravaged Galveston in 2008, my 91-year-old grandmother moved about thirty miles away to safer ground near Clear Lake, Texas, while my aunt decided to remain in Galveston and rebuild her severely flooded home. Under the new construction code Galveston adopted after the storm, homes in her neighborhood had to be built high enough to survive the level of Ike's storm surge, which was about twenty feet. My aunt's home was the first structure in the neighborhood to be completed and when it was, it towered outlandishly over every other home you could see. Standing on her porch, I was looking down on most of the other houses' roofs. Such is the shape of things now if you are going to be ready for Galveston's next storm.

The new building code was only one of the changes made to prepare Galveston for the future. Ike's path and devastation bore a striking resemblance to the 1900 Storm, and the response to Ike is also sounding familiar. Local leaders have developed a plan to build up barriers to protect a vast swath of the Upper Gulf Coast centered on Galveston, a plan dubbed the "Ike Dike." This project would be in many ways a twenty-first-century version of Galveston's original seawall but covering much more territory. The proposal includes a set of massive 800-feet-wide steel barriers that would swing closed across the mouth of Galveston Bay when a storm was coming. The rest of the Ike Dike would rely upon engineered sand dunes and other raised barriers, as well as Galveston's existing seawall, to cover a seventy-mile swath.

Ike certainly raised awareness about the vulnerabilities of the Houston/Galveston region, and recent studies have modeled worst-case scenarios should a much stronger storm make landfall near Galveston again. An extensive review of these studies by the *Texas Tribune* and ProPublica in 2016 forecast damages far beyond

Ike, with a particular focus on the highly industrialized and strategically significant petrochemical complex along the Houston Ship Channel. This study imagined a storm with nearly 150-mph winds and thirty feet of storm surge. Hundreds of thousands of Houston homes would be flooded by such a storm, some up to the second and even third floors, and coastal areas like Galveston would be completely flooded even before the storm made landfall.[2] The industrial and economic damages are projected to be astronomical. Consider that about a quarter of all gasoline and two-thirds of jet fuel, and an unknown proportion of the US military jet fuel, are produced by oil refineries along the Ship Channel. In addition, more than 150 chemical plants line the Ship Channel, producing about 40 percent of the nation's basic chemical inputs. The economic disruption that would result from these facilities going offline for an extended period is difficult to calculate. While the estimated cost of the Ike Dike has grown from $8 billion to $30 billion and beyond, these massive figures pale in comparison to the $100 billion-plus in losses that such a future storm could bring.[3]

As both a native of Galveston and an expert in renewable energy, I can't help but ponder the utility of the heroics that would be involved in executing a project like the Ike Dike, if it is not paired with meaningful engagement to confront the drivers of climate change and the obstacles to our effective response I already outlined. Bill Merrell, the individual who first came up with the notion of the Ike Dike, said about his idea: "The concept is easy. You stop the storm surge at the coast so that you protect everyone."[4] But what about those living along the hundreds of miles of the Gulf Coast outside its protective walls? In fact, an Ike Dike creates a risk that displaced water that would have flooded homes in Galveston partly surges, with even greater strength, around the barrier's ends into adjacent communities, damaging homes there. And what about the stronger and more frequent forest fires ravaging other parts of the country, and the rising seas, spreading tropical diseases and other harmful changes coming? How many protective walls can we build for all the threats we face?

It is telling that among the items being debated in Ike Dike planning is whether to size barriers for a storm as severe as Ike or for an even stronger storm. Remember, Ike was only a Category 2 hurricane, on a scale extending up to Category 5. Given that meteorologists are now debating, in light of unusually powerful recent storms, whether to add a new Category 6 to the scale,[5] it seems that there is a strong possibility that the Ike Dike may well end up being another opportunity to outsmart ourselves—a way to think we have protected ourselves that paves the way for delaying the hard choices we know we must make to reduce the greenhouse gases that are causing the storms in the first place. One estimate is that planning and funding for the Ike Dike might be complete by 2025 and construction may be finished by 2035, when I'll be nearly seventy years old.[6] It will be my children's generation that will have to worry about all the other Ike Dikes that would be undertaken to defend against climate change elsewhere around the country.

In contrast, my youth in Galveston was spent looking up at the stars and wondering what incredible things we might do next. My father was a Navy Frogman, today known as a SEAL, and he led the Navy's team collecting astronauts upon splashdown. He told me how he flew out to the capsules in a helicopter, opened the hatch, and transferred the astronauts back to dry land. Later, he volunteered for another space-related mission, TekTite II, which placed scientists in underwater living quarters as part of an experiment to investigate space travel. I had a photograph of my father with Gemini astronauts in my room and marveled at his stories about living in an underwater spaceship. On television I watched *The Jetsons* and *Johnny Quest* and at the movies, *Planet of the Apes*, *Star Wars*, *E.T.*, and *The Right Stuff*. For me, the future was a place of hope and dreams—and this is what seems at stake in our current climate crisis. For younger people today and future generations to come, the changes coming are painting a darkening picture, compounded by current leaders' failure to meaningfully engage the fight. As a parent and, perhaps one day, a grandparent, I find myself during my less hopeful moments trying to limit my emotional attachment to only a small universe of my potential future family—to my children and their children, but none after that—as a way to care less about the fate of the world, should we fail to take up this fight in a timely manner.

There is another risk to Galveston's seawall and the Ike Dike that may be built—one far beyond the question of how much it may cost, how many feet high it should be, or how it should be configured. This risk has to do with rising seas and the question of, I suppose, whether the dike is being planned for the right place. Recent studies forecast that, depending on exactly how quickly melting of the Antarctic ice sheets and Greenland's glaciers proceeds, it is possible that Galveston's entire seawall, along with any Ike Dike and of course the city sitting behind it and communities up and down the mainland behind that, will be completely submerged under rising seas by the year 2100.[7]

While they don't have catchy names yet, projects similar to the Ike Dike are being considered for New York, Boston, New Orleans, Miami, and other major US cities.[8] In New Orleans' case, they have now begun studying a massive new barrier project[9] because the project they just finished, the ten-year, $14 billion effort undertaken in the wake of Katrina to rebuild levees and pumps sufficient to protect the city from the next storm, will no longer offer that protection starting in 2023.[10]

Walls are not the answer, at least not the only answer, when it comes to keeping the ocean at bay. At some point, we must address head on the things we are doing to warm the planet.

■

What will become of us?

It is helpful to consider the long view, given the length of the story of climate change heretofore, and whatever is to come over the years, decades, and centuries ahead. In this regard, one per-spective comes from *Big History*, a fascinating book written by Cynthia Stokes Brown in 2012, which posits that the way we teach history to our children is all wrong. The world didn't begin with Mesopotamia and the Greeks and end with September 11, as my kids' textbooks seem to suggest today. Rather, it started with the Big Bang and it extends not just up to the present but also through the known facts we have about what the future holds. Brown's insight—that we need to broaden our perspective significantly to understand the whole story, so that we can see our own plight, for better and for worse, in its complete context—illuminates a num-ber of surprising aspects of our situation, but one rises above the others: the centrality of climate to human history.

Big History points out a fact that is pretty obvious, once you think about it: that all of human civilization's "greatest hits"—from the invention of agriculture through the rise of nation-states, written alphabets, religion, the atomic bomb, and the iPad, and almost everything else you can think of—have all occurred in something of a brief geologic time period of climate anomaly, which Brown calls the "10,000 warm years." Most of us misper-ceive the climate of these warm millennia as simply the way the world has always been and will always be in the future—an obvi-ous, and grave, error apparent to anyone with a rudimentary understanding of geology. This way of thinking is perpetuated by our widespread classroom myopia. History books from ele-mentary through high school practically never venture outside of these 10,000 years. Brown argues that, through this pedagogic blind spot, we misunderstand the precariousness of our species' connection to the climate in which we were incubated and then came to thrive all across the globe. This blind spot is our enabler in the opening of the Pandora's box of climate change via green-house gas emissions.

Other books echo *Big History*'s perspective on the very long tra-jectory of our relationship with energy, extending over millennia behind us and ahead. Among these are *Children of the Sun* by Alfred Crosby, *The Birth of the Anthropocene* by Jeremy Davies, and *The Collapse of Western Civilization,* Naomi Oreskes' and Erik Conway's dark, science-based fictional work. *Children of the Sun* presents a brief and insightful high-level telling of the whole energy story of our species, and *The Birth of the Anthropocene* recounts the academic debate around the placement of climate change in the geologic time scale. *The Collapse of Western Civilization* is a fictional report set in 2393 upon the third anniversary of 2093's imagined "Great Collapse" of modern civilization, precipitated by climate change. That the authors of *Collapse* are not science fiction or horror genre authors but rather a pair of science professors at Harvard and the California Institute of Technology tips the reader off to the deep factual grounding of the fictional account set forth. If you are interested in climate, energy, and public policy but can't spare more than a few hours to read something more on this topic, you should put the short but explosive *Collapse* high on your list.

Together, these books illuminate how choices we have made over the past few centuries have already converged to define a surprisingly narrow range of outcomes for humankind. It is becoming increasingly clear that regarding energy and climate, our species has roamed the broad plain for many millennia, but over the past two hundred years or so we have been barreling forward into a narrowing canyon.

These works provide a unique perspective on the question of what climate change's most potent and consequential effect may be. Across most of the world, the human story has been understood as one of soaring achievement transcendent far beyond our lowly evolutionary roots. We have been a species that has dreamt audaciously and that has defined itself in pursuit of our tremendous aspirations. We have undertaken lofty ideals, fought great wars, invented fantastic technologies, and prevailed in all manner of grand struggles, all throughout our history and up to the present: traveling to the moon, conquering subatomic science, inventing medicines and life-saving technologies that exceed hard boundaries of only a handful of years before, not to mention the daily miracle that is the global machinery of feeding, educating, and nurturing billions of us in our far-flung corners of the world. We surely haven't won all of our battles, but those we haven't, we mainly just haven't won yet. Most importantly, we hold this viewpoint of the ascendant arc of our story as powerfully about our past as we do about our future. We understand ourselves within this mythology; we see in ourselves an essentially godly agency, that we can achieve just about whatever we decide to, sooner or later—because, by and large, we have always done so. This is our human story, a story perhaps especially strongly held here in the United States as a result of our truly exceptional history.

To me, climate change's gravest threat is that it will upend this transcendently optimistic human narrative and transform it completely. Unquestionably, climate change carries the capacity to rewrite the human story from an epic of soaring achievement to a colossal, self-inflicted tragedy. So much of what we believe about ourselves, our families, our nations, and our species is at odds with the idea that the world we encounter will deteriorate—rather than improve—with each generation. This is, after all, the very essence of the American dream we are talking about.

All this helps to finally put the promise of solar energy and the remarkable story of how it came about fully into perspective. The Barilla plant demonstrates that photovoltaic solar can now stand on its own two feet, doing battle with fossil fuel power plants and prevailing against them on their home turf: utility wholesale markets. This is an extraordinary, historic moment, all the more so given how our electricity markets today bestow a price advantage on fossil fuels by ignoring the costs of their carbon emissions. This achievement alone points the way to ever-expanding zero-emissions energy sources coming to serve a growing portion of our energy needs, displacing fossil fuel power plants and diminishing greenhouse gas emissions on a rolling basis. As an economic force unfolding in the market, even with no change in carbon policy, it represents the bend in the arc, the beginning of the end of the Combustion Age that we initiated early in the eighteenth century, and a start on the long, green road back to renewable energy. If we develop the political will to correct our markets so that they may begin addressing the known effects of greenhouse gas emissions, Barilla's economic force will be multiplied and joined together with other existing technologies, and with new ones yet to be developed, to become a thundering, stampeding force for good. From the long view of big history, solar energy's maturation—encapsulated in the little Barilla Solar power plant in a remote town in West Texas—contains the seed of a powerful turning point in human history.

Two aspects of Barilla's story reveal its power to change our current path in fundamental and positive ways. One is the fact that Barilla is the result of two centuries of cooperation—messy though it was, entirely unplanned at times and only partially intended at others—among the world's countries, each contributing its respective strengths to advance the common agenda of cheap, ubiquitous clean energy. It is not difficult to imagine that there is a time on the horizon when cheap and entirely clean energy is so plentiful that it is available not just to elevate our quality of life across the globe but also to power the new ideas not dreamed of yet that will address the daunting problems its predecessor technologies generated. Barilla and the remarkable collaboration around the world that created it represent just one example of the transformative results we can achieve if we work together in our response to the existential threats that face us. Likewise, solar energy's astonishing history of collaboration demonstrates what is at stake in a world in which cooperation deteriorates and the nations of the world decide to "go it alone." If enormous technological challenges must be solved to manage the multiple threats on the horizon, it is obvious that steps to foster international collaboration will be vital to meeting them.

The other lesson Barilla has for us is the most promising and potentially transformational of all: the idea that the stunning small project in West Texas represents the appearance of global capitalism as a beneficial force with incipient capacity to radically reduce worldwide greenhouse gas emissions around the world, instead of exponentially multiplying them. That our massive and ever-growing appetite for power, the engine of human activity everywhere we wander, might turn from driver of catastrophe to an inhibiting force and protector against it represents something few might have dreamed we would ever see—a transformative rebirth of global capitalism, perhaps occurring just in time.

Preventing the worst effects of climate change is a solvable problem that the world will either fully undertake to address in time, or not. In my estimation, history books will record the decades between 1980 and 2030 as the half-century in which we knew what was happening and what we had to do but struggled with the idea of taking real action. Much of this period has already passed with no coalescing of the requisite political will to engage, as demonstrated most recently by the woefully inadequate Paris Accord and, here at home, by the Trump administration's decision to exit from it and his numerous policy decisions harmful to serious engagement.

At this moment, the nature of our ultimate response is still very much up in the air. There are powerful forces of atomization that would pull the nations of the world apart, and communities within them—the Trump threat, writ large. Donald Trump was, at best, the wrong man to be president at the wrong time; at worst, his "every man for himself" philosophy on climate change is the embodiment of humanity's clay feet when it comes to responding to a challenge of this scale. His four years in office have passed, but Trumpism as a political philosophy of betting against each other, rather than on us all, succeeding when the going gets tough, is out of the bottle: if climate change is a grave threat to life on the planet as we know it, then Trumpism is the equally grave threat to our ability to collectively respond effectively to it.

Despite the soaring rhetoric of the post-war period about global cooperation and world peace that underlies international institutions like the United Nations, the question hangs very much in the balance: Will we join together around the world as we sometimes have to overcome existential threats, or will we divide ourselves into conflict and strife? If every country proclaims its own interests will come "first" before the plight of all the rest, we place ourselves in a definitively reactive posture as tragedies unfold, unquestionably leaving us weaker in addressing the root causes of the problem than we would be if we were united. Even worse, with nations and communities within nations turned against each other and fighting over dwindling resources instead of collaborating to forge solutions, the promise of Barilla, and all the future Barillas to come, would recede from our grasp. Without strong and growing international collaboration, the path to reducing greenhouse gas emissions in the quantities needed and along a quick enough timeline is virtually impossible to conceive.

There are also unifying forces at work, of course, appealing to our better selves to attack the problem with our best and brightest and enlisting collective action from across the globe. The election of Joe Biden and his inclusive, multilateral approach to climate progress is a big step back towards the right path, although only time will tell how successful he will be in passing the kind of legislation needed to implement his plans. This path puts us in an active posture to formulate the rapid, decisive, and muscular response so obviously called for, allowing us a fighting chance of ending up in the range of better possible outcomes.

Every nation has a role to play in determining and executing our response as a species, but it is very clear that we Americans can, should, and must play a special role. After all, more than any other single country, we have created this problem, not just through off-the-charts greenhouse gas emissions we produce year upon year but also through our devotion to the real engine behind those emissions: capitalism. Wide-scale adoption of fossil fuel use may have started in England with the Industrial Revolution, and China may be its most prolific convert lately, but American capitalism, which we have successfully evangelized around the world for well over a century, has undeniably been the most powerful multiplier of greenhouse gas emissions worldwide.

I believe America will play this role. But first, we Americans must decide to engage.

■

When will America rise to the challenge to take up the fight in earnest? This is the question of the hour when it comes to how humanity will respond to climate change.

It is said that when Winston Churchill peered into America's soul in the run-up to World War II—when we were dithering over whether or not to pile into the conflict engulfing England and the rest of the free world—he observed, "You can depend upon the Americans to do the right thing, but only after they have exhausted every other possibility."[11]

This could be an observation about human nature, about all peoples, everywhere, hesitating in the last moments before deciding whether or not to commit ourselves to some massive, epic undertaking—to the hardest path among the many before us. It is a sensible thing to do, after all: exploring all the alternatives in order to be certain that the hardest of paths is indeed the right one.

But with regard to World War II and the titanic undertaking then at issue, and to the unique position the United States occupied with its ability to strongly influence the final outcome, Churchill's insight was certainly about a specifically American dilemma. There was no other nation on earth that could have turned the tide to win the war by joining the battle, and surely the burden of this knowledge lay heavily in the American mind and heart. This had to have been particularly difficult because, at the time, there was both little appetite for undertaking to resolve a mainly European conflict and a certainty about the enormous loss of life that would be required to prevail—hence, the prolonged dithering. Surely there was a way to avoid this crushing endeavor. Surely there was some shortcut. Could we just lend and lease mountains of equipment to our allies and watch them carry the day from afar? Could we merely provide transport, goods, funds, and moral support, and call that enough?

Polls on the eve of American involvement in the war showed where we stood, and they confirm Churchill's assertion. We knew the war was coming, and we supported taking steps to prepare for it, but we resolutely opposed actually getting into it. In 1940, 62 percent of Americans believed that entering the war to defeat Germany was the right thing to do, and 89 percent supported introduction of the first peacetime draft in US history. But at the same time, only 29 percent thought that we should declare war on Germany and send troops into Europe.[12] This very nuanced position reflected Churchill's assessment exactly: We knew what was right and also that we would do it, but up until the last moment we would do anything but start down this path.

It seems obvious that Churchill's insight into the American character following World War II is also on point with regard to climate change today. Americans know the right thing to do; we understand our responsibility to act; we believe our country will play a unique role in confronting the problem; and yet, we are reluctant to fully engage and commit ourselves. It is no coincidence that Americans' views on climate change uncannily reflect the nuanced views we held in 1940 on the eve of World War II. Sixty-five percent of Americans believe that climate change is caused by human activity, 64 percent worry about it a great deal or a fair amount, and 75 percent believe it is a threat to the country. However, at the same time only 31 percent believe "immediate action" should be taken.[13]

For all but the relatively small number of Americans alive today who lived through the conflict, American exceptionalism and preeminence seem to be a given. When we reflect today upon the saga leading up to World War II and its end, as well as prior and subsequent major threats that we have overcome, we feel that it was a foregone conclusion that we would not only join those fights but also that we would prevail because we are, unquestionably, the most exceptional country. Even if it took us a bit of time, of course we beat the Axis powers and then reshaped most of the world as we saw fit; Churchill merely put into words what always has been and always will be true about us Americans.

There is much that is proper and right about this uniquely American view of our own history, and much that defines our quintessentially American mythology. But with regard to the latest threat to rise up against us—the incipient crisis of climate change—there is something quite important here that we are overlooking. The missing piece is the period of time in the calm before the storm in which we questioned ourselves, our place in the world, and whether or not we would take up the fight—the time when we stared deep down into the abyss and struggled with whether we truly *would* find the grit and wisdom to do the right thing and carry the day.

The matter of how the bad news piles up to create a grave period of fear, reflection, and self-doubt, and then transforms somehow into resolve, and finally action; the question of whether it really is inevitable that we Americans will do the right thing; the crucible in which it is revealed whether we are the nation the world believes us to be, the people whom we tell the world we are and whom we believe ourselves to be—these are at the center of where America is now on climate change and where the world will go from here.

We Americans have done this before, and we prevailed in spectacular fashion. Not only did we prevail in the existential fight that threatened us in the 1940s, but after our victory we undertook to recast much of the world into an architecture of uniquely American vision. This vision has its faults, to be sure, but it forged a world order of unprecedented peace and progress all around the globe: the Pax Americana, our greatest gift to the world so far.

It is hard to see why the threat of climate change could not provide the same opportunity to put the world on a new track that adapts and mitigates climate change and finds the means to transform the world's vast CO_2 and other emissions infrastructure into an ever-cleaner resource.

At the end of the day, as many reasons as there are for grave pessimism as I write this late in 2020, I am ultimately an optimist about our chances. I do not doubt that Americans will, eventually, come together and lead the world to meet this challenge. We do not help ourselves by forgetting our history and the comfort it provides: that each time prior crises have seemed hopeless, and victory seemed unlikely—if not outright unimaginable—Americans rose to incredible heights to prevail. Only our timeliness in coming together is in question, then as now, and on this we must hope that Churchill was right about us—that we will not let the crucial last moment to act pass us by.

A new dawn has risen with the Biden administration and its quick action to rejoin the Paris Accord and resume so many other sensible policies to address climate change. For many of us, the prior administration's antithetical approach to climate matters made it harder than ever to believe that we will rise to this challenge. As much as I would like to believe it is not so, the 2020 election results—Joe Biden's narrow margin of victory and an evenly split US Senate—do not entirely alter this picture. However, anyone cognizant of the full scale of climate change and its trajectory in time need not be overly concerned about the pendency of any single American presidency. Trump is now gone from the White House and Biden has begun his term, but climate change has been hundreds of years in the making and is now unmistakably, at best, a hundreds-years more crisis only in its infancy. As terrible as Trump's presidency was for climate change, and much else, and as vexing as steadfast Republican opposition to Biden administration climate efforts may prove to be, Americans may take meaningful action on their own, as Europe, China, India, and numerous other countries have done—even while the sleeping giant that is a committed America continues to procrastinate.

The assertion that neither a very active Biden administration nor Donald Trump's four-year hiatus on climate action will be the be-all and end-all on climate change may sound like heresy to some, but it is folly to believe that the presidency is our main problem. Consider Biden's narrow victory, in full view of the Republican Party's policy vacuum on climate matters, and that four years earlier, had Trump lost in 2016, Hillary Clinton would have faced the very same Republican majorities in both houses of Congress firmly opposed to taking just about any meaningful step to address climate change. Consider also the significant but ultimately limited advances that the Obama administration was able to make to combat climate change during its eight years, sparring with the same kind of uncooperative Congress during a majority of that time. The growth in installed renewable energy achieved during the Obama administration was historic but nonetheless massively inadequate relative to the emissions reductions needed. Sadly, much would suggest that President Biden's efforts in a similarly divided government will succeed in the same kind of incremental advances, but not the transformational progress the moment requires.

Ultimately, one cannot deny that Trump and his ilk are merely a symptom of America's lack of engagement and resolve on climate policy, not its cause. The Biden administration has hit the ground running on climate change, picking up where the Obama administration left off, but the 2024 and 2028 elections are not far away and will be consequential on climate matters. Regardless of which candidates run in those elections, Trumpism and the notion that we still need not take climate change seriously will certainly be in the mix. Only when the country makes up its mind to take bold action will it reject Trumpist candidates and the governors, senators, and representatives who espouse such views. And only then will the threat they collectively represent be vanquished.

It is helpful to recall that the resiliency of American democracy has perhaps been its greatest strength. Without exception, during the run-up to the existential crises in our past we have suffered through our worst presidents, while the moments of transformational engagement that followed have seen our greatest. For every Franklin Pierce and James Buchanan, respectively ranked by historians forty-first and forty-third worst out of forty-three presidents, there was next an Abraham Lincoln, ranked first; for every Calvin Coolidge and Herbert Hoover, ranked twenty-seventh and thirty-sixth, respectively, there came a Franklin Roosevelt, ranked third.[14] George W. Bush was ranked thirty-third worst, and one may only conjecture where Donald Trump may end up in history, but in many ways the answer seems clear already. I have full faith that, in this pattern among American presidencies, we have now lived through those in the former category and may surely expect the arrival of the latter. It is possible that they already have: while the obstacles to doing so are formidable, President Biden and Vice President Harris may find ways to transcend the politics of the moment and rouse the nation from its slumber as our truly great leaders did against all odds in the past.

It is also useful to note that when it comes to grave issues like climate change, the US presidency may well not end up being the most important global leadership position on the matter. Of the ten most influential figures in American history, as judged by historians, only five were presidents.[15] Likewise, the accomplishments of world leaders galvanizing the kind of modern social movements that the global climate struggle will require—those guided by such towering figures as Mahatma Gandhi, Martin Luther King Jr., and Nelson Mandela—have paralleled, if not vastly exceeded, those of the most accomplished elected leaders around the world. The scale and complexity of undertaking the challenge of climate change may well be better suited to leaders capable of devoting a more persistent, lengthy, and focused effort than any political office accommodates.

Our time of self-doubt and vacillation has long expired. The call of those who distract, disbelieve, and deny the obvious facts visible to our own eyes has gone shrill. The time is unquestionably now to take a stand and begin meaningful action on the scale required.

What will become of us? There are many reasons to fear, but there are many more to start believing in ourselves and take action. At the end of the day, we must acknowledge that the only clear answer to this question is that our fate lies entirely within our own hands, and the sooner we start the hard work, the better.

EPILOGUE

THE WORLD THAT MAY BE

Imagining a future America in which we have acted to transform capitalism to address climate change.

[Author's note: While some things about the world imagined below may seem unlikely or even outlandish, I have extrapolated each event, invention, and geopolitical occurrence from recent news. To learn more about the real-world bases for the predictions, see this chapter's endnotes for expository information.]

The year is 2049. You are flying from Los Angeles to New York City for a meeting. You survive the takeoff; boy, you sure miss the old days when it was jet engines that eased you into the sky. The discovery of lightweight battery chemistry in the late 2020s revolutionized air travel, among many other things, making flights everywhere quicker, if also a G-force adventure. For an old-timer like you, even worse than the takeoff is the descent, when all but two of the plane's propellers reverse direction so they recharge the batteries all the way down.[1]

As you round over the Pacific out of LAX, you spy the new double rows of seawater desalination plants lined up on land leased in front of the expanded base at Huntington Beach Naval Station. What a story that was—the president, Pentagon, EPA, governor, and mayor all working together to balance the myriad federal, state, and local concerns to complete these massive projects. Finally, Angelenos, and everyone else in Southern California, are assured of a plentiful *water supply for the first time in as long as anyone can remember, and quite an affordable supply, too, since dirt-cheap electricity is its main ingredient.*[2]

Heading east now, you survey the megalopolis that is, today, greater Los Angeles. The city has begun a slow, careful march, literally picking up many buildings and infrastructure and carefully relocating them east, just ahead of rising tides and storm surges advancing year on year.[3] *Back in 2020, the region was home to about twenty million souls, but in the two-and-a-half decades since it has almost doubled in population. Yet the air is cleaner than it has been in generations, and the economy is thriving. No one foresaw the growth back then, given the water shortages, fires, and escalating temperatures, but in retrospect, neither California's boom as the world's undisputed leader in clean energy nor the flow of refugees from points south into the relative order and safety of American cities should have surprised anyone.*[4]

Silvery reflections from the rooftops below twinkle in the afternoon sunshine; you recognize the solar paint that is today ubiquitous everywhere there are people.[5] *Thanks to the massive technology push of the 2020s to reinvent solar, and then reinvent it again, and again, a sprayable solar paint is now cheaper, stronger, and longer lasting than house paint. Some version or other of this technology now coats just about every manmade surface that stands still long enough to be sprayed—not just rooftops but also whole skyscrapers, highways, parking lots, and sports stadiums. Once applied, not only does the stuff pump out electrons, even on an overcast day, it also reflects back into space much of the sunlight it can't convert into electricity. Cities used to be heat islands, five to ten degrees warmer than rural areas, but now they are almost the same temperature as the latter—which is saying something, given the brutal heat waves that increasingly park themselves over much of the country.*

As greater Los Angeles gives way to the desert, the first of the solar generation stations begin to appear, neatly organized into discrete projects here and there in order to preserve wildlife habitat and open space between them. The federal government's anachronistic ownership of nearly half of all western US lands,[6] previously a source of controversy when ranching and other uses were at least somewhat viable here, has been transformed into our potent arsenal of clean-energy generation. Who knew that these largely desolate tracts of public lands extending in all directions, devoid of fossil fuels and unusable today for just about any other economic purpose, would be so rich in the very resource we are most in need of today?

These islands of solar projects extend from the Mojave across the entire sunny southwest—through Nevada, Arizona, Utah, New Mexico, western Colorado, Texas, and beyond. Each panel ultimately connects to the fully integrated national network of massive direct current power lines marching to urban centers east and west, north and south, all the way to both coasts. The entire country is now linked in a single, balanced grid supplementing population centers with the power they do not produce themselves. Just like railroads used to carry coal from Wyoming and Pennsylvania to power plants all over the country, these powerful transmission lines now carry clean electrons from the many renewable resource areas to all the load centers.

As you leave California, you look left out the window at Las Vegas, off in the distance. The city is struggling. Hot days there, which used to reach about 105 degrees in the summer, have been regularly hitting 115 and 120, and the all-time record went from 117 way back in 2017 to 133 degrees last year. Before the solar network was completed, the only power plant that dependably worked on the hottest days was the Hoover Dam plant, but when Lake Mead pretty much dried up, even Hoover couldn't be counted on any longer.[7] The very first blackouts gave the casino business a black eye, and then the following year real estate prices surprised everyone by simply collapsing. People have been moving out ever since, and today Vegas is getting close to being one of many ghost towns dotting the American West.

Phoenix and many other large southwest cities have similar, if less extreme, stories.[8] Ironically, today each of these cities has access to more energy generation than it ever had before to meet the growing need of managing hot temperatures—but just when the power problem began improving, the water problem became intractable. With a diminishing water supply and the hazardous heat spikes making outdoors off-limits much of the summer, life in the desert cities is finally becoming just too hostile.

About when you reach Kansas, large clumps of wind turbines begin to dot the landscape, organized, like the solar stations, into discrete groups to maximize open space. The turbines churn in unison across the broad plains where the winds reliably rip from north to south, all the way from Canada to the Gulf of Mexico. Unlike the old style of turbines, which had their seventy-ton generators sitting atop tall steel towers, these drop the generator to the ground, reducing costs by an order of magnitude—one of the many changes wind turbine manufacturers had to make to survive the onslaught of cheap and plentiful solar power.

If these wind projects were rocks in a stream, you could hop from one to the next, from Canada's Northwest Territories across Alberta and Saskatchewan, down through the Dakotas, across to Montana, and Wyoming, then through the Great Plains, and all the way down Central Mexico to Oaxaca, and then the mountains of Costa Rica, one of the windiest places in the world. Last night, while you were sleeping—with California's solar projects all awaiting the dawn—power from this multitude of interconnected wind projects charged the batteries in the airplane you are in now, as well as all the batteries in the massive energy storage farms parked outside of each city from coast to coast. With these storage facilities offering plenty of power to bridge gaps when the wind and sun aren't generating, the grid is at least as resilient as it ever was, even though a growing majority of power comes from these variable sources.

Crossing next into Kentucky and Ohio, every now and then you spy the towering smokestack or cooling tower of a massive old coal, natural gas, or nuclear power plant. These are mostly curiosities now, laid low as cheap and clean generation came to dominate the same power markets fossil fuel plants had once ruled. Many of the plants that are still standing are completely abandoned, but some cooling towers have been turned into bungee jumps,[9] and yet others have been transformed into museums, extolling the great accomplishments of the Combustion Age and the multitude of incredible feats that fossil fuels made possible, both in the United States and around the world. It was a different time, now in our past, but a time in which this great country was built and first came to project its power far and wide.

Of course, fossil fuel generation is not gone completely. Large-scale, ultra high-efficiency natural gas power plants remain in operation here and there across the country. Although they are rarely called on to operate, when they are needed they reliably meet whatever demand appears. It's helpful to have these sources of energy there when we need them, which will be from time to time for some time yet, as our dependence on electricity for all aspects of our life has never been greater.

As you reach the Ohio state line, you look left and can just make out Cleveland in the distance. Cleveland's skyline has been an unmistakable sight since the 125-story Apple iPower Tower was built not too far from the 100-story Amazon Power building and just down the street from the Shell Power & Light tower. All are, not coincidentally, easily accessible from the Tesla Tube Midwest station, conveniently located right downtown.[10] Not long ago, Cleveland would have been an unusual choice for such important corporate headquarters, but lately no one prefers the coasts, desert areas, or forest fire zones. As oceans rise, the Great Lakes have receded some, so there is no threat of inundation there either.[11] Cities like Cleveland, Minneapolis, Buffalo, and a resurgent Detroit all used to be flyover country but now are on the cutting edge.

Apple, Amazon, and Shell got into the electricity business in the 2020s, along with Exxon, Google, Walmart, and Tesla.[12] This began with mass protests demanding climate action around the same time. What followed next was surprising, at least at the time. When the

Republican Party was finally pressed by its own voters, its opposition to serious climate action suddenly collapsed. In the end, the Republican leadership's transformation was not unlike an earlier case of sudden political reversal: the evolution of Republican leaders on the issue of gay marriage back in the early 2000s, when increasingly shrill opposition gave way almost overnight to a dramatic realignment with the party's reasonable majority.[13] *Deeply out of step with their constituents, Republican politicians were ultimately led from behind; rank and file members gave their leaders the freedom to support sensible climate policies, or the boot if they didn't.*

This change in politics led to landmark legislation fundamentally reforming power markets, which in turn caused a sea change in the American electricity business. Reforms, led by a coalition of climate hawk Democrats and market solutions-oriented Republicans, imposed a robust carbon tax on fossil fuel generators and wiped away all vestiges of the monopolistic constraints favoring big utility companies like Allegheny Energy and American Electric Power.[14] *Effectively transforming fossil fuel power plants overnight from "assets" into "liabilities" on their ledgers, these reforms were the last straw for many incumbent utilities that had steadfastly resisted the renewables revolution. Goldman Sachs was the first to take a short position on Southern Company—an unbelievable occurrence for this perennial blue-chip stock—but other investors quickly followed suit. The contagion rapidly spread to other old-school utilities in a merciless bloodbath calculated purely on carbon exposure. The entire sector was quickly remapped as Big Tech and an already green-trending Big Oil rushed in to fill the void.*

Oddly enough, all this progress arose out of the turmoil of the 2020s. The 2028 election turned out to be Trumpism's last stand and, after the dust cleared, the Republican Party, the nation, and the world breathed a sigh of relief. Americans clearly rejected the go-it-alone, "nationalist" mantra as counterproductive, if not outright harmful, to the country's interests. For the first time, both parties in the 2028 election were speaking of climate change as more of an economic opportunity, national security, and social justice matter than an environmental issue, and the message resonated so broadly with voters that both parties came to Washington on inauguration day with fresh ideas about how to proceed.

The new administration plowed ahead with sensible policies that attacked emissions as aggressively as they promoted wealth equality and environmental equity. In foreign policy, the international community of nations was stitched back together, grounded in America's restored credibility on the world stage. This work, paired with observable progress on policy targets and economic growth, inspired people to believe that climate change was not such a lost cause.

Even more promising was the new administration's work to partner with private businesses, particularly Big Tech, Big Oil, and the major defense contractors, to search for new approaches and solutions for addressing the multiplying challenges facing the nation. These efforts commenced with a critical research mega-project to identify safe, dependable, and effective technologies for removing key greenhouse gases from the atmosphere in large enough quantities to begin to move back, gently and safely, the climate change dial. The inclusion of machine learning and artificial intelligence in these efforts will, it is hoped, accelerate the timeline from the original estimate of a century or more to merely several decades.[15]

Nearing New York City now, your plane pivots above the Atlantic in preparation for landing at LaGuardia. As it banks, you make out the wave machines swaying in ocean swells below you, the latest innovation in what is now called "low-consequence" energy technologies.[16] Thinking back on the three hundred years of near-exclusive fealty to "high-consequence" combustion power sources, it's hard not to wonder at the colossal lack of creativity we once demonstrated in meeting our energy needs. Despite being a curious, inventive, and resourceful species, we all but ignored the massive quantities of energy naturally surrounding us, long past the time when we had become aware of the harm we were hard-wiring into our future.

Off to your left as you descend above the East River is the United Nations headquarters. The organization has done great work since its existential crisis of the late 2020s, when an array of key members threatened to withdraw and the group entered into a period of insolvency and chaos. It took a major effort by the United States and other nations for the UN to reemerge, but not before the plenary membership definitively addressed the matter of its fundamental inability to achieve meaningful results on climate issues, with cascading consequences for member countries around the globe.

From that terrible crisis came the most important and far-reaching policy the UN ever implemented in its work to fight climate change. It dedicated most of its 3 percent allotment of worldwide carbon tax proceeds to the innovative Fragile States Reinforcement program, which is helping to sustain a number of nations in crisis and to invest in their communities to bring opportunity to many millions of vulnerable citizens. This initiative is also mitigating the drivers of the international refugee flows that have imperiled the most vulnerable populations worldwide and destabilized areas in Asia, the Middle East, Africa, Europe, and Latin America. Fueled by multiple, simultaneous crises of famine, civil and regional conflicts, extreme weather, and natural disasters, a phalanx of failing governments in these regions had transmitted a contagion of violence across borders in much the same way that the West, in its Cold War "domino theory" a century before, had imagined communism would spread from nation to nation.[17]

As you come in for a landing, flying hovercraft clear out from in front of you and the city streets become easily visible. You can pick out "gas stations" selling both types of fuel—not leaded or unleaded gasoline, of course, but green hydrogen and electricity—and grocery stores here and there with their distinctive science-fiction look, characterized by large, bulbous, glass greenhouses extending out over the streets. Inside each one a wide variety of vegetables grow on multiple levels, maintained with careful attention to temperature, nutrients, pH, etc.[18] Grocery store hydroponics are just one of the changes that made perfect sense as the effects of the carbon tax rolled through the economy; tomatoes trucked in from upstate, or out west, definitely taste better, but they are now a lot more expensive. Most of the produce people buy nowadays comes from these greenhouses; the days when California provided almost half of the nation's produce are long gone, since the massive Central Valley infrastructure projects of the early twentieth century finally ran out of water.[19]

As you walk through the airport, troops in camo walk by in groups or sit waiting to board. It's a sad fact that our troops are more or less omnipresent these days. The world is aflame with conflict, and the United States often finds itself in the middle of it. EU and UN troops support us on many missions, but neither of the most populous countries in the world—India and China—are much help. This is not surprising, though, given that they are occupied by internal conflicts and, perhaps inevitably, by a looming conflict between them: the so-called "Water Wars" to determine who will control the dwindling supply.[20]

As you exit the airport and climb into an autocab—apart from the occasional airplane flight you're a landlubber, no aerocab for you—you reflect on how quickly the world has changed in so many ways. It wasn't that long ago that people were complaining there were no flying cars. Also gone are the days of burning fossil fuels to take two tons of steel with you to the neighborhood grocery store and back. You still see combustion engine vehicles on the street from time to time—mainly driven by conspiracy theorists and the AI-resisting Luddite community—but fewer every year. Gasoline engines have been banned entirely in most major cities, at least outside of the United States.

At last, you arrive at your destination. It's a hot, sunny day—not that unusual anymore for November in New York. All long-term predictions are for temperatures rising even further over the rest of this century and for at least a couple more—but, perhaps, not more than that. If emissions keep falling all the way to zero by 2060, as planned, and if the AI greenhouse gases removal project succeeds, the *best science available tells us that we will stabilize at an average temperature around 4 degrees Celsius hotter than that of 1980. As great as it is to have temperature stabilization very possibly within reach, it won't mean the climate will stop changing any time soon. Even a successful CO_2 removal technology would likely take a century or more to move the dial on the quantity now present in the atmosphere, and the warmed air and oceans will keep melting glaciers and ice sheets for an unknown period, flooding more coastlines.*[21]

And who knows what else is going to pop up in the intervening years? For instance, the planet has been experiencing an unexpected temperature elevation over the past decade. The reason is uncertain, but climate scientists suspect it is caused by reductions in particulate pollution—smog—from all the fossil fuel power plants having closed down. With less and less smog shading the planet, more sunlight penetrates. How ironic: first we made the planet hotter by polluting it, and now we are making it even hotter by ceasing to pollute it.[22]

In the meantime, due to advances in scientists' understanding of ice sheet dynamics, the exact rate of sea level rise across every coastal region around the world is now predicted through 2100 to within about five years. You look up the flooding forecast on Google for the building you are now walking into and find that this very entryway will be breached in about sixteen years, barring something unexpected.

Block by block, the inundation timeline of every seaside neighborhood, not just in New York City but also in Miami, Shanghai, Lagos, and Rio, has been published for all to see. Paradoxically, disclosure of this information has calmed stock prices of the highly exposed real estate, insurance, and infrastructure companies, and just in time. Many thought making the inundation schedules public would throw these corporations into financial turmoil, but it turns out that predictability and transparency were all investors needed to process the changes that are coming.

Climate scientists can provide a higher degree of certainty now than ever before about what further changes are coming, and when, but it still feels bewildering. The "unusual" has become usual, and very unusual things seem to happen all the time. Few things seem like they should, but then again that makes sense, since the whole world is changing in ways no one has seen before.

Still, somehow, you are hopeful, and even optimistic. As they say, the first thing to do when you find yourself in a hole is to stop digging, and we have not only done that, we have also started to fill the hole back up. It feels like the country is now behind the effort to do what needs to be done, and so is much of the rest of the world. To be sure, this is no kumbaya moment—there is plenty of argument about exactly what should be done, some of it strident—but the days of each side hating the other went away when denial finally died. The vast majority of Americans are fully engaged in the fight; the debate today is about how to most effectively reduce and then reverse emissions, not whether to do so.

For those of us who remember the good old days of the twentieth century, we know things won't ever be the same as they were then. But that's okay. We were living on borrowed time then, time we knew then was borrowed from today and the years to come. We're paying it back now and making it right, paving the way for a new future in which our place on the planet is secure. Optimism has returned, even though times are worse around much of the world. Now we can once again dream and strive like we did in the days when we thought there was no limit to what we could do and no problem we couldn't solve. If only we had focused on solving this one earlier.

AFTERWORD

TURNING THE SHIP IN THE RISING STORM

As climate reports pile up since the first edition of this book was published documenting ever-closer tipping points—and some already past—the pace of humanity's response gathers momentum, even as the ways we may fall short become more apparent.

A s I sit writing this in Austin, Texas, in September 2023, out my window an increasingly familiar Texas summer weather pattern grinds on: a merciless high-pressure dome drives record-breaking heat and dry conditions day after day, week after week, month after month, on repeat like the *Groundhog Day* movie. We have already seen a tripling of our average of fifteen days over 100°F per year, and the forecast offers no sign of relief. And Texas hasn't been alone this summer in experiencing extreme heat: blistering temperatures have tormented most of the lower Northern Hemisphere, from California to Florida, and across Northern Africa, Southern Europe, the Middle East, Central India, and much of China.

There's nothing unusual about it being hot in Texas and many other places in August, of course, but even so, something feels different about this summer. While many of us have become numb to news of historically hot conditions—like that the past nine years are among the ten warmest ever recorded—this summer, the sheer bulk of truly extraordinary heat conditions around the globe makes it seem that a page is being turned now in front of our very eyes, from a normal, if unusually hot, summer, to something definitively different.

The meteorological case for 2023 as a turning point in future histories of our era is already strong, even with several months left to go. The year began on track to be a very hot one with March the second hottest ever and May the third hottest, but then came the summer. June, July, and August were not just the hottest ever recorded, they were the hottest by a very significant margin.[1] In the context of the steady and consistent annual warming increases over the last twenty years, it is all but mathematically certain that 2023 will be the hottest year ever and by a wide margin, sticking out of the annual temperature charts like a hockey stick.

The example of Phoenix, Arizona, provides a window to what this extraordinary heat meant in a place many Americans know well. Phoenix is used to excessive heat, but its breathtaking string of fifty-five days over 110°F this summer beat its 2020 record and, more notably, burst the prior record, from 1974, not by 10 percent or even 50 percent, but by more than 350 percent.[2] The city's off-the-charts July record of 102.7°F mean temperature, which included record overnight lows as high as 97°F, made it the first major American city to have a monthly average temperature over 100°F.[3]

As extreme as this summer heat was, we can at least take comfort in the fact that higher temperatures reflect the predictions of the UN Intergovernmental Panel on Climate Change (IPCC) climate reports, the gold standard in climate forecasting. These reports, which were most recently updated over 2021–2023, represent the most current and thorough information we have on the changing climate picture.[4] These new reports update the rate at which global temperatures are expected to climb to an even faster pace than the prior reports, leading *Scientific American* to print the sobering headline this August, "This Hot Summer Is One of the Coolest of the Rest of Our Lives."[5] At least when we see the IPCC forecasts coming true, we can feel we have a handle on what is coming—but the reverse is true as well, that climate events falling outside the new forecasts portend we are hurtling into uncharted territory.

This summer brought a couple of such concerning events. The first was the soaring ocean temperatures that broke out abruptly around the world beginning in June. Average ocean temperatures have historically varied only by small increments year to year, but this summer about half of oceans worldwide experienced marine heatwave conditions, meaning temperatures are above the 90th percentile for the prior thirty years,[6] and many areas had temperatures so extreme that they literally shot off the historical temperature charts.[7] Ocean temperatures off Florida have rarely been far from their average 88°F, but this summer averaged more than five degrees higher, and occasionally reached into the high 90s over several days in July.[8]

Scientists have struggled for an explanation of the sudden and extraordinary marine warming, with some speculating that climate models may have incorrectly assumed a linear warming trend for oceans, when in fact, perhaps beyond some threshold, the rate of warming may become more exponential in nature.[9] Since oceans have absorbed the vast majority of greenhouse gas–driven warming that the earth has experienced so far, this development raises the worrisome prospect that we may be facing, much earlier than anticipated, limits to the oceans' ability to moderate future surface warming.

The other extraordinary event that has similarly alarmed climate scientists this summer was, strangely, technically a winter event. Beginning in July and extending through August, a completely unprecedented extreme heatwave swept across the Southern Cone countries, including Argentina, Chile, Paraguay, Brazil, and Uruguay, right in the heart of their winter season. A bizarre high-pressure heat dome spread across the region, bringing temperatures well over 90°F and in some places over 100°F, breaking records for winter highs by a mile. In North America, the equivalent event would be Virginia or

Kansas experiencing days of 90–100°F, but in the middle of February.

Scientists assess that the culprit behind these recent extreme warming increases is likely one or all of three recent developments. First is the 2022 eruption of Hunga Tonga-Hunga Ha'apai volcano near Tonga, which blasted a vast amount of water vapor, a potent greenhouse gas, into the upper atmosphere. Second is normal occasional increases in energy the earth receives from the sun.[10] Third is the rapid shift of Pacific Ocean dynamics that began this spring with the change from La Niña to El Niño.[11] An El Niño weather pattern is characterized by very warm conditions across most of the equatorial Pacific, which, in turn, drive generally warmer weather in most of the rest of the globe. This year's El Niño came on strong as the summer began and appears to be exacerbating already warmed surface and sea conditions, conceivably driving the anomalies presented above, and much more. This theory certainly seems convincing but, if true, it raises yet more concerns, both about future El Niños and about the current one as it extends into 2024.

A recent study by the Chinese Academy of Sciences' Institute of Atmospheric Physics peered into the future of the current El Niño event and forecasts a likelihood that a strong El Niño pattern will continue through the remainder of 2023 and into 2024. The investigators' model indicates that a record quantity of heat has accumulated in the upper ocean levels during this El Niño, heat that in 2024 could drive "severe consequences for the global ocean," including marine heatwave intensification, ocean deoxygenation, and thermal expansion. Because warmer water takes up more space than cooler water does, 2024 ocean warming could be expected to worsen sea level rise, storm surges, and saltwater intrusion into coastal freshwater sources. While the report makes clear that it is not yet certain that 2023's prodigious El Niño will continue into 2024, it finds a 60 percent chance that it will mark a new record for El Niño warmth and set the stage for a 2024 comparable to or more extreme than 2023.[12]

Alongside these record conditions has been an exceptional laundry list of serious political and economic challenges. The world has emerged from COVID-19 precautions into societies changed in numerous, complex, and significant ways, some for the better and others worse. The Biden administration has begun to put its stamp on domestic and world affairs, leading a clean break from most Trump policies while also, surprisingly, continuing others; Russia's invasion of Ukraine resuscitated Cold War dynamics and inflamed global tensions across a range of critical matters; and, perhaps most important, the United States and China seem to have commenced a new era of hardened decoupling and bristling tension, a development with grave potential to alter world events for decades to come, on climate and much else.

Making serious progress on emissions even while the world continues to turn with its non-climate challenges and opportunities has always meant we must learn to multitask—to "walk and chew gum at the same time," as my father used to say—to stay on track with both, even as the two categories increasingly intermingle in often vexing ways. Global refugee flows, for instance, are endemic in human history but will be increasingly driven by fused vectors like domestic and regional conflicts underpinned by climate crises, such as when drought or flooding creates food and water insecurities, which, in turn, catalyze armed conflicts and populations fleeing for safety from them. These types of dual challenges will no doubt increase in the future, although they appear to be materializing more quickly than we have thought.

The news on the root driver of the many climate changes we're seeing—that is, our global annual greenhouse gas emissions—includes some bright spots but overall continues to be very negative. On the positive side, global solar and wind installations are soaring in 2023 by the largest increase ever, 107 gigawatts, to a

total of 440 gigawatts of new additions, an amount equivalent to
the installed power capacity of Germany and Spain combined, and
on to 550 gigawatts in 2024.[13] China is leading the way, accounting for more than half of new wind and solar installations, dwarf-
ing Europe and the United States. These are staggering quantities
of new renewables entering into operations each year, and more
important they are the result of a gargantuan scale of manufacturing capability put in place all around the world in recent years that
will churn out more equipment available for deployment. Each of
these new projects will displace fossil fuel power plants that would
otherwise have been constructed and further drive into retirement
old and uncompetitive coal, oil, and gas combustion plants, often
the most potent greenhouse gas–producing plants there are.

On the negative side, a web of complex events and factors are
continuing to drive emissions upwards well in excess of Paris
Agreement commitment levels and at a pace forecast to cause
global average temperatures to exceed dangerous "point of no
return" levels sooner than previously forecast. The global economic slowdown caused by the COVID-19 pandemic lowered
emissions briefly, but since then three factors—the worldwide
economic resurgence following the pandemic, the loss of Russian
natural gas imports into Europe caused by the Ukraine war, and
increasing extreme weather events—have all propelled large new
greenhouse gas emissions.

Asia has seen the biggest growth in emissions lately, particularly
from coal, which has returned to such an extent that it is overshad-
owing significant advances the region made building solar and
wind. China saw crippling power shortages during its record heatwave and drought in 2022, and then experienced record power
demand growth after lifting COVID-19 restrictions. Faced with a
choice of waiting for more renewables and energy storage to come
online or green-lighting additional coal plants, China prioritized
economic growth over their Paris Agreement commitments. The
country has increased coal generation by 14 percent since 2022

and recently approved 106 gigawatts of new coal plants,[14] representing about two new Californias' worth of new generation and associated emissions. Even as China pushes forward with its ambitious solar and wind energy plans, today the country burns more coal each month than the rest of the world combined.[15]

A similar dynamic has affected India, which experienced the hottest March since 1901 in 2022 followed by the hottest April since 1901 in 2023. These and other extreme heatwaves exacerbated endemic power shortages in the country, which threaten economic development and continuing efforts to lift many more millions out of poverty. India—today the most populous nation on earth with 1.4 billion people and the world's fifth largest economy—has launched major solar and wind development around the country but still relies upon coal for about 75 percent of its power generation.[16] While India's Paris Agreement commitments always foresaw a longer time period to curtail emissions than other major emitter nations, recent actions seriously call into question India's recent new commitment to reduce coal to no more than 50 percent of its total generation by 2030. Instead, coal use has increased the last few years, driving up 2022 emissions by 15 percent,[17] with more on the way: currently more than 32 gigawatts of new coal plants,[18] the equivalent of about one new Georgia in generation and associated emissions, are under construction in India. The country is projected to surpass the European Union in total annual emissions soon.

Sadly, it is not just Asia backsliding on its emissions commitments lately—Europe and the United States are slipping, too. When Russia's two Nord Stream natural gas pipelines into Europe exploded under the Baltic Sea in 2022, longtime climate stalwart Germany did something very surprising in response to power shortages: it brought several coal plants out of retirement and reopened its lignite mines, increasing its coal consumption by 20 percent over 2021.[19] Lignite is one of the dirtiest resources to mine, easily mistaken for ordinary dirt, and the most carbon-intensive type of coal. Recent data suggests that emissions from increased coal use in Germany and other similarly situated European countries as a result of the loss of cheap Russian natural gas may be offset by new renewables additions there, and also by decreased energy consumption due to high electricity prices. However, if the war in Ukraine continues for some time, which certainly seems possible today, then coal's revival in Europe could as well.

The upshot of coal's recent global resurgence is that while its use had previously peaked in 2013, it set new records in 2021 and 2022, and the answer to the question of when it will decline again remains uncertain. Even more consequential is the fact that each new coal power plant built has a thirty-five-year expected lifespan, laying the groundwork for all these new plants continuing to complicate climate goals through the decades. Similar circumstances have driven recent global oil consumption such that forecasts for peak oil demand continue to be pushed out: the International Energy Agency forecasts oil demand to rise by 6 percent between 2022 and 2028, with demand reductions from electric vehicle adoption to hopefully begin to counteract strong growth from petrochemicals and aviation by the end of the decade.

In the United States, we continue to see coal generation and our overall emissions decline and renewables quickly increase, although in our case, cheap domestic natural gas from fracking has been coal's greatest market nemesis. Nonetheless, as the world's largest historical greenhouse gas emitter by a mile and the second largest today, our success in dramatically reducing emissions as quickly as possible is critical. We remain off track on our Paris Agreement commitments, and forecasts suggest that even with recent changes, including the Inflation Reduction Act (discussed in more detail below), we will not meet our 2030 emissions goals or the later ones.[20] Like those of other developed countries, US emissions increased with the post-pandemic recovery, and it is not clear yet when they will begin to fall again and then plummet, as we have committed they will.

The recent Paris Agreement backsliding and returns to coal in Asia and Europe suggest that a troubling pattern is emerging: when push comes to shove and countries are forced to decide whether to prioritize economic growth or their clean-energy goals, many are choosing the former. In politics, near-term benefits often come at the expense of long-term sacrifices. But the impact of these backward steps on the imperative to rapidly reduce emissions worldwide is grave. In April 2022, UN Secretary General António Guterres was painfully direct in his comments: "This is a climate emergency. . . . The science is clear: to keep the 1.5°C limit agreed in Paris within reach, we need to cut global emissions by 45 per cent this decade. But current climate pledges would mean a 14 per cent increase in emissions. And most major emitters are not taking the steps needed to fulfill even these inadequate promises."[21] As if Mother Nature saw fit to emphasize Guterres's statement, not even one year after his call to arms, for the first time ever global average temperatures exceeded the Paris Agreement's 2030 maximum temperature limit of 1.5 Celsius above pre-industrial levels—before dipping below again.

These points relate to the critical matter of how we generate the power we use, but the other side of the equation, how much our consumption of power is growing, is similarly important. Growth in demand for energy worldwide is forecast to be nothing less than staggering, particularly in electricity use. The International Energy Agency finds that global electricity demand will rise somewhere between 5,900 terawatt hours (TWh) and 7,000 TWh by 2030—a quantity of new consumption equivalent to adding *a new European Union and a new United States* on top of existing demand.[22] Drivers for this increase range from changes that may reduce overall emissions, such as growth in electric vehicle use which will diminish gasoline combustion, to those likely to drive far greater emissions, such as rising air conditioning demand across the many areas experiencing warmer temperatures.[23]

Some new changes on the world scene have yet to be assessed, including the arrival of high-powered AI-based computing systems, such as ChatGPT, which are expected to become widely available across homes, schools, and businesses as we seek to utilize the many benefits of these powerful new technologies. While there is great hope for the positive changes AI technologies may help us achieve, including in the area of climate change, some fear that exponential growth of AI technology utilization is already assured, and, along with it, massive new greenhouse gas emissions. AI applications rely intensively on cloud computing, which already drives incredible growth in power consumption. These new emissions would come on top of all already forecasted.[24]

How can one make sense of how all these positive and negative factors balance out in the larger picture of climate change in 2023? One way is to return to the simple metric presented in chapter 2, the carbon budget. This concept provides a means of understanding how many tons in greenhouse gases we can emit in the future before we are certain to exceed the quantities that will cause us to exceed the 1.5°C and 2.0°C warming levels beyond which scientists believe extremely negative changes are virtually certain to occur. The carbon budget was updated in the new IPCC reports issued recently,[25] and the most recent research finds we have already consumed nearly all of the 1.5°C carbon budget and will pass that threshold in the next couple of years. The 2.0°C budget will more likely than not be passed later this decade.[26]

As I write this afterword, the question of where all these momentous and challenging occurrences will leave us weighs heavily on me. In the period since I started writing the first edition of this book, our three children have gone from middle and high school kids to two college students and a recent graduate. As they each now enter adulthood, the contrast between the future that prior generations contemplated and the one that young adults like them face today is stark, and they feel it keenly.

A 2021 study published in *The Lancet* assessed 10,000 youth aged 16–25 around the world, including both developed countries like the US, France, and Australia and developing countries like India, Brazil, and Nigeria. The study found that 60 percent were "very" or "extremely" worried about climate change, 76 percent felt the "future is frightening," 56 percent felt climate change makes them think "Humanity is doomed," 52 percent felt their family security will be threatened, and 39 percent described themselves as "hesitant to have children."[27] The majorities were similar across the countries surveyed, although generally more extreme positions were expressed by participants in developing countries.

I see these results in my own children and their friends. They describe to me a foreboding of a degraded future, alongside a desire to get involved and do something about it, whether, in the case of my two younger children, in the field of environmental work or, for my oldest who majored in film, in sharing stories about what our future may hold and how we think about the challenges we will face. I am heartened that among the actions each of my kids is very motivated to take is to vote, a viewpoint that seems to be generational—recent trends in youth voting suggest turnout rates are significantly on the rise.[28]

These results suggest that for most youth today, no matter where on the planet they live, the outlook is challenging. From where I sit, there is reason both for their alarm and to take heart that changes are occurring—although they remain incremental in nature, not nearly the transformational pivot that will be required.

■

As the story of global emissions has charted its upward course over the last few years, I am fortunate to have had a front-row seat to a separate trajectory—that of the power business unfolding as a result of the "Energy Transition" discussed in chapter 4.

While working on this book, I started a company with several friends in 2017 that focuses on large-scale, power-plant-sized energy storage projects. The central idea of this business was one at the core of the Energy Transition: that as more of our electricity comes to be generated by solar and wind, which create cheap, clean power, but only when the sun shines or wind blows, and less of it from coal and natural gas, which is more expensive and creates emissions but can generate any time needed, power grids will experience bigger fluctuations in supply that will have to be reconciled with constant, and growing, demand. Solar and wind have many benefits, but people also still want the lights to come on whether or not it is windy or sunny at that moment. So, our idea for this new company was to create new power plants that could help match ever-shifting supply with continuous demand by building projects that store power when there is too much of it on part of the grid—such as on a windy or sunny day in West Texas—and then discharge that extra power later when it isn't so windy or sunny. As the number of solar and wind plants grows, we believed, so would the usefulness of this new kind of power plant.

Fortunately for us, due to incredible advances in the same kinds of lithium-ion battery technologies that run our phones and laptops, there is now a way to build these new kinds of energy storage power plants. Lithium-ion batteries were invented by scientist John Goodenough at the University of Texas and colleagues, earning him and two others the Nobel Prize in chemistry in 2019, and since then electric vehicle–focused companies invested large sums into R&D to make lithium-ion batteries last longer, weigh less, store more, and operate safely for years at low degradation rates. All this progress made possible the first grid-connected battery energy storage projects during the 2010s, and they have been increasing ever since.

Power-plant-scale lithium-ion energy storage technologies have come along at just the right time to match the incredible shift in how power grids driven by the Energy Transition operate—a change that is likely even more fundamental in nature than most think.

Historically, grids came into being in a very logical way. As the first power plants were built in the late nineteenth and early twentieth centuries, they were constructed as you would expect in close proximity to the population centers they served—which placed the vast majority of them in or near our large cities. Over time as cities grew, more power plants were built, and transmission lines were built connecting them and extending service out to suburban and then rural customers. As these networks grew large enough geographically to come into contact with each other, transmission interconnections were built between them, a system of rules was adopted to govern the systems, and they became the early versions of the grid systems we have today, all but one of them, Texas, now interconnecting to each other across all of North America to help when one system has too much or too little power for their needs.

The rise of renewables that is described in this book has been turning this all on its head. What has been happening as the Energy Transition has come on the scene in the late 1990s has been a slow-motion inversion of the grid system that existed before. Wind and solar projects have been the fastest growing types of new power plants built over the past few decades, and they fit best in remote, rural areas with enough land to match their much less dense energy resource needs. This has meant that, increasingly, new generation is being built far from the population centers that consume most power, while old coal and natural gas power plants near our cities have been unable to compete and are closing down. So, while the location of where most power is consumed has remained the same, the location of the generation providing that power is changing with great acceleration away from cities and toward remote areas.

The upshot of this "Great Inversion" of the grid has been to set the stage for a reshuffling of the values of services provided by grid-market participants. Previously, generators provided power reliably delivered when and where most customers consumed it, which made their product very valuable. After the Great Inversion, more and more generators are remote from their customers and, furthermore, they must rely on long transmission lines to reach them, lines built to ferry vast amounts of power from the cities out to small rural communities, not the other way around. These rural power lines were much smaller and quickly met their limits on windy or sunny days, creating traffic jams of electrons that keep them from reaching customers, an eventuality that can compound the natural intermittency of renewables. In this evolving grid, many new power generators are unable to provide power at all the times customers want it and must therefore find other solutions to fill those gaps. In this new market, value is shifting from generators to those that can manage these new, increasing gaps in time and location for customers, and energy storage is an excellent way to meet these needs.

Managing these gaps was the opportunity our new energy storage company was focused on, but when we started to look for investment in 2018, we were surprised to find that many investors very familiar with solar and wind were nonetheless nervous about investing in energy storage. The risks of storage, they felt, were high owing to the fact that battery technologies were new and markets had not developed contracts for buying services from them yet. We expanded our search and eventually came across different investors more open to the types of risks we faced—investors more comfortable with taking risks on new technologies and market opportunities. Ironically, these investors came not from the renewables world but from the oil and gas sector. Fossil fuel investors quickly understood and became comfortable with the risks our business would take on, which to them seemed not very different from risks they took in the oil field—and perhaps also because they were looking to hedge risks of their own, like that of seeing the Energy Transition reduce demand for fossil fuels, leaving them high and dry. Our new business plan offered these investors an opportunity to diversify into a risk profile they were comfortable with, operating in a sector they wanted exposure to.

We closed on the investment that launched our energy storage company in the fall of 2019 from a private equity group that had theretofore only invested in oil and gas ventures. Three years later, this company, Jupiter Power, had constructed six large-scale, energy storage projects operating every day in power markets to meet the kinds of needs our business plan had intended: projects ranging from 10 megawatts to 200 megawatts in size we used to absorb power when it is plentiful and discharge it when it is scarce or when frequency regulation is needed. In 2022, this private equity group exited their investment in Jupiter via sale to another, much larger private equity investor, also one with limited prior exposure to Energy Transition investments. Today with this group's backing, Jupiter continues to build new storage projects and to plan many, even larger ones to serve the expanding need for services helping to smooth out the growing proportion of power provided by solar and wind generation—which continue to show no signs of slowing down.

The story of Jupiter Power's start and growth—including its backing by institutional private equity investors with limited prior focus on low-carbon ventures—is a microcosm of how private institutional capital is getting behind the Energy Transition these days and, beyond that, of the new ways we are actually engaging, for better and for worse, in funding the fight against climate change. The forces driving this transition, as discussed in chapter 4, are durable and continue gaining momentum with each day, and it is no surprise that capital providers far beyond those focusing solely on "green" investments are increasingly involved. Putting money to work on cleaner energy is no longer a decision of "Should we take a loss but do what is good for the environment?" It is more often these days, "We need to diversify beyond fossil fuels, or we could potentially lose money," or even, "We need to become a leader in Energy Transition investing if we want to be competitive."

Jupiter's story also sheds light on two larger truths about the

Energy Transition. First: to the extent a massive share of the emissions driving climate change is the power plants that fuel our world, climate change can be seen as not just an environmental, ecological, and social problem but also very significantly a problem of infrastructure—essentially, the challenge to rebuild, all around the world, the ways we get the energy we need but from sources that produce little or no emissions. If it is so that climate change is significantly an infrastructure problem, and if infrastructure problems are problems requiring access to plentiful and cheap capital to address these issues, then by the transitive property, a big part of addressing the problem of climate change is about getting access to much more and much cheaper capital. The addition of new private equity investors to increase the supply of capital to get this infrastructure built is a positive step forward.

A second truth: the participation by major private equity and institutional investors in the fight against climate change draws attention to the relative absence of another major investor, arguably the biggest source of cheap infrastructure capital of them all—the United States federal government. Whereas throughout history major infrastructure and technology challenges have almost as a rule seen the federal government play a substantial and direct role, in the case of clean energy addressing climate change we are seeing very little. The biggest federal support we have seen is the Inflation Reduction Act (discussed below), but it provides no federal funds for direct investment. This approach stands in contrast to historical examples where federal funds played direct roles in addressing the nation's big infrastructure problems, such as the interstate highway system, the Los Angeles Aqueduct, the Chickamauga and Hoover dams, and much more. Some have noted this limited approach of the US and some other governments to climate change and questioned whether we are seeing a shift in the way we deal with large problems generally.[29] Whereas Franklin Roosevelt and other US leaders often deployed large-scale public funds to manage major policy problems, today's approach seems instead to be to leave those opportunities for major private investors. Will the federal government eventually enter the fray directly, or are we witnessing the "privatization" of major public challenges to institutional investors?

A closely related concern, some would say, is that big private equity investors who have profited for decades from activities that increased the harmful emissions driving our climate problems should not now be allowed to profit from the business of cleaning up that mess. Private equity writ large has been a fundamental part of the investment community that has provided oil, gas, and coal players the funds they have needed to grow, and they have benefited with gargantuan profits from these activities over the last several decades. Some efforts to assess the carbon footprints of the largest funds have concluded that in a few cases they rival those of industrialized countries.[30]

Should companies with records like these be trusted to now become part of the solution? From my perspective, there is no question that the answer is yes. It is, in fact, crucial that those investors who reaped profits from carbon-intensive investments now plow those gains into solutions and swiftly begin putting their very formidable capital and talents to work addressing greenhouse gas emissions. My perspective is a pragmatic view, and reasonable minds may certainly differ on this question, but the problem of climate change is so vast and so dire that we need all the resources we can bring to bear in addressing it, and it is encouraging now to see fossil sector energy investors moving swiftly and meaningfully into clean energy. It is now common to see private equity groups touting their growing clean-energy investments and accomplishments proudly on their websites and advertising this work in front of employment recruits, many of whom highly prioritize positive-impact climate work as a part of their career paths.

When I recently attended a meeting with our private equity investor in their New York City offices, I stepped past a throng of climate-action demonstrators to enter the building, but once inside, each presentation I attended detailed for investors in the fund the different efforts the company is undertaking to accelerate renewables and other clean-energy investments.

It is worth pointing out, too, that the list of investors who profited from emissions-creating activities does not just include private equity funds—far from it. Fossil fuel exploitation over the last century and more has been one of the greatest investment opportunities in history, and it had no shortage of interest from investors of all stripes. Pension funds providing crucial income for hundreds of millions of retirees, sovereign funds underwriting government budgets, and banks, insurers, and other institutional investors that underwrite business transactions around the globe each day join perhaps an even larger historical block of fossil fuel investors—hundreds of millions of individuals putting their savings into mutual funds and stocks of companies that have driven and in many cases continue to drive greenhouse gas emissions, including the likes of ExxonMobil, General Motors, Facebook, Amazon, and Google.

For all these reasons, from the perspective of someone who has toiled in the renewables industry for more than twenty-five years and often struggled to raise all the capital we could profitably put to work, the turn towards clean energy by major new investors, like private equity groups and others who formerly invested in emissions-producing activities, is a very positive sign of the sea change occurring in the investment community. Each of their investments in Jupiter, and in the numerous and growing number of other Energy Transition ventures, contributes to the solution—and simultaneously reduces the capital available to fossil fuel investment opportunities. This sea change is already delivering vastly more capital to the kinds of projects that will continue to drive our global emissions down and seems poised only to grow in size.

■

Much of this book—one could even say, the core thesis of this book—is about the idea that humanity can achieve incredible things when we focus our resources on a hard problem and work together to solve it. The scale of the problems we face from climate change will require a broad and potent response to be effective, and the times through history we have responded effectively to challenges of great scale have arrived only when we have come together under common cause. Climate change is a unique stripe in the great pile of problems of massive scale in human history—it is perhaps of the greatest scale, affecting all of humanity across the entire globe and all its future generations—and so gathering those affected into common cause, focusing global wherewithal on the steps required to reach solution, and inspiring and leading all to action will be especially formidable tasks, and tasks that at their core require *working together with our smartest and most capable from all over the world* to reach that common goal. The story told in this book of solar energy's incredible advances that led to the Barilla project and its progeny of cheap, clean gigawatts appearing each year around the world is one in its essence about climate progress achieved through countries working together, and in particular about the United States, Europe, and China working together. Unfortunately, the global state of affairs has lately been tilting in the wrong direction on this front, on climate change and much else.

If we could see a Venn diagram illustrating the shaded area that is the overlap of Donald Trump's policy agenda and President Joe Biden's, most would be surprised to learn that within that slender lens of agreement lies the continuation of the tariff war with China that the Trump administration initiated. The vast majority of tariffs the US imposed on Chinese goods being imported into the United States, and the Chinese tariffs imposed on US goods being imported into China, have remained in place. Also in the overlap is an embrace of a much bigger US policy change with regard to China—we have seen a continuation of the Trump administration's escalation of tensions between the two countries. US policymakers rightfully and appropriately protest Chinese human rights abuses, its persecution of Uyghurs in Xinjiang, its military expansion and aggressive actions against the US and our friends in the South China Sea, its systemic deployment of unfair trade practices and currency manipulation, its corporate espionage, and much more. China for its part objects to US policy on Taiwan and human rights, US trade practices, including particularly restrictions on technology exports, and legal action taken against Chinese companies, like Huawei and TikTok. Overall, it seems that Trump and Biden agree that the era of a policy of "engagement" with China as the best way to secure US interests and objectives has ended, and a new era of confrontation has begun.

From a climate point of view, this development introduces the potential to significantly alter the trajectory of how we will, as a species, approach the challenge of climate change. The United States and China are the two largest emitters of greenhouse gases, and the European Union, which is broadly aligned with the US regarding China, is close behind, and the soft power of these giants to push the rest of the world in a positive climate direction is without parallel, meaning that as go the total future emissions of these three parties, so goes the story of climate change and humanity. The most effective tools we have developed so far—cheap and ever cheaper ways of generating plentiful electricity without creating net new emissions—are the fruit of the collaboration of these three, so each is linked fundamentally not only by its complicity in causing climate change but also by its cooperation in addressing it.

As noted above, if we are to respond effectively to the climate challenge facing us, we must learn to walk and chew gum at the same time with climate issues and all the others with which the world must deal. The United States, China, and European Union must find ways to work together and with the developing countries continue to advance climate progress effectively and collaboratively even as we confront the problems between us, including serious ones. It would be productive to explicitly memorialize an agreement among these three world powers to separate, to the maximum extent possible, matters of progressing emissions reductions from all other affairs. A next step would then be to extend this coordination agreement to a larger group, the G20 countries—several members of which will be highly ascendant in importance to climate matters over the remainder of the century, such as Brazil, India, Indonesia, Russia, and Saudi Arabia. In the absence of such an agreement to segregate and prioritize continuing climate cooperation, nations of the world may continue to drift in the direction that the Trump administration blazed its trail on climate policy—each country "going it alone" and eschewing cooperation and collaboration.

The most consequential event to have happened recently to strengthen climate progress—the United States passing the Inflation Reduction Act (IRA)—is an epic achievement, but it is also to a degree a manifestation of the "go it alone" approach to climate that began with the Trump presidency. President Biden has his own rationale for leaning this direction; in his case, it does not seem to be mainly a China animus, as it seemed to be with Donald Trump, but rather the goal of growing domestic manufacturing jobs that underpinned the law's "America First" undertones.

The Inflation Reduction Act is a game-changing law passed by Congress and signed by President Biden in 2022 which, contrary to the law's name, is fundamentally about putting the United States on a path to rapidly advancing the Energy Transition in the United States from coast to coast. Its key provisions include extending important tax credits for solar and wind for a decade, transitioning to a technology neutral clean-energy tax credit starting in 2026 and continuing them until CO_2 emissions reductions targets are met, creating new incentives for emergent technologies like hydrogen, carbon capture, and energy storage, linking incentives to prevailing wages and training/apprentice requirements, providing lifeline income to certain nuclear power plants, and incentivizing new investment in disadvantaged energy communities, closed coal mines, and power plants.

Perhaps the most important components of the IRA, however, are those that promote domestic manufacturing of key Energy Transition equipment, with the objective of wresting manufacturing jobs from China and other countries that would otherwise be exporting these items to the US. The law establishes significant and ongoing incentives not only for the construction and operation of manufacturing facilities to make clean-energy equipment but also for procuring the key strategic commodities required for that manufacturing. These provisions cover how to obtain resources necessary for the Energy Transition, like rare earth metals, lithium, and much more, from within the US and from friendly nations, like Australia, Canada, Chile, Mexico, and others, on an advantaged basis. These changes not only create a basis for a return of manufacturing jobs across the country tied to the success of the Energy Transition but also establish a means of improving access to strategic minerals and de-risking the supply chain overall for renewables and other clean energy–related equipment.

The potential impact of the IRA on the growth of renewables, energy storage, and emergent clean-energy technologies is hard to overestimate. In addition to creating millions of new jobs, some forecasts call for annual US renewables and storage installations to double within just the next few years, and then increase again by the same amount by 2030, resulting in exactly the kind of grid replacement scale of new construction needed to catch up to our Paris Agreement commitments. Others believe the build forecasts are so extraordinary, and the razor-thin Democratic majority that achieved passage of the bill so small, that its successes may be short-lived pending changes in control of Congress and the presidency. Indeed, Republicans already attempted, unsuccessfully, to repeal key IRA components in the 2023 debt ceiling deal and have promised to attempt a full repeal in the future. Time will tell, but the key pieces of an architecture supporting a robust climate response fueled by private investment have been put into place by the IRA.

As positive a development as the IRA is for the United States's climate goals, the law's "America First" preferences for equipment and key commodities point to a potential new challenge to emissions progress by inviting other countries to respond in kind. The more each country insists on favoring its own equipment and inputs over others, the further out of reach the benefits of capitalism's core tenet of specialization become, and the more expensive these critical weapons in the fight to mitigate and adapt to climate change will be. In other words, if every country decides to create preferences for equipment and materials from their own countries and put in place barriers like tariffs on such goods from other countries, the kind of collaboration among nations that led to the Barilla project—with the United States, Europe, and China each bringing their respective strengths to bear to make renewables as cheap as possible—risks falling out of reach. China has managed to produce the cheapest solar and battery equipment in the world, standing upon the shoulders of American and European scientists and entrepreneurs who invented and advanced the original core technologies; from a climate change point of view, it would be very counterproductive for a cascade of anti-imports laws to hinder outcomes that most effectively advance our emissions reductions goals.

In chapter 10, I lay out my suggestions for actions we must take to effectively confront climate change. In light of the factors discussed here diminishing the effectiveness of the world's major powers working together on climate change, I add an additional recommendation: *It is time that major countries of the world agree to eliminate tariffs on mature clean energy equipment.*

What are tariffs and how on earth could they be important to climate change? Tariffs are taxes that a government puts in place on targeted goods imported from another country in order to protect domestic producers of the targeted goods or punish the exporting country, or both. By making the imported good more expensive for domestic customers, the imported product becomes less appealing, which benefits domestic sellers of the good, to the extent they can undercut the new higher import price.

Tomes have been written on the efficacy and economic theory of tariffs and whether they efficiently achieve their objectives, but it is not necessary to go into a dry discussion of these factors to make the point that if we are serious about addressing climate change, then it is time for tariffs on these goods to go—and this is especially true in a world where each country adopts some version of the IRA for itself.

The biggest reason to end these tariffs now is that, insofar as cheap renewables constitute the most potent tool humanity has developed to date to reduce the emissions driving climate change, public policies like tariffs that operate to make these products *more expensive* directly undermine the highest-level climate policy directives of the very governments implementing them—not to mention regular old common sense. If we want solar panels and wind turbines and other proven emissions-reducing equipment to become more widespread, and if we believe in market economy capitalism, then the last thing we should be doing is making said equipment cost more—we need to be making it cheaper.

Eliminating tariffs is a logical step in a world where more countries are incentivizing clean energy with subsidies as the IRA does. If the objective of IRA subsidies for domestic manufacturers is to help them compete with foreign imports, then we do not want to artificially inflate imported products by tariffs because their lower prices will set the market for the domestic producers. It should be one or the other, subsidies or tariffs, or else end-use customers will be paying unnecessarily high prices.

In addition, both tariffs and subsidies distort market prices and can create unintended, highly inefficient outcomes. For example, customers buying foreign solar panels are paying the tariff to their government as tax revenue, and then pass that increased cost along in the form of higher energy prices from that solar panel over its entire lifetime. Then, to take taxpayer money to fund incentives for solar and wind projects that subsidize a project's all-in costs, including the tariffs paid in the case of imported equipment, puts the government in the position of simultaneously taxing and subsidizing the very same item. Project owners are paying higher taxes to the government via the tariff and then getting tax incentives from the government that they, in turn, use—to subsidize the higher prices caused by the tariff! These conflicting market distortions create a pointless "do-loop" inefficiency from a tax policy point of view that is at odds with itself.

A couple of final points are worth emphasizing regarding my recommendation to end certain tariffs. One is that ending tariffs on Chinese-manufactured solar panels should not mean also ending the essential and appropriate ban we have put in place on all Chinese imports made using slave labor. Fighting climate change is critically important, but a future that tolerates slavery and mistreatment of minority communities, such as Uyghurs, is not one worth securing.

Another is that any agreement for limiting tariffs should only cover mature technologies, like current technology photovoltaic solar panels and wind turbines. Tariffs should remain available for new innovations in order to keep countries honest on fair-trade rules and to allow governments to protect the advantages of new inventions that their research and development investments yield.

■

As I conclude this afterword, I will return to the turn of the twentieth century and events that happened then that we are only now learning to see in the proper way.

Science has a way of usefully casting what we thought we understood in a completely different light. In 1907, as Galveston was completing construction of the first phase of its seawall and grade raising, across the Atlantic another epic lesson in man's hubris regarding Mother Nature was just beginning. Over dinner, two London shipping industry executives agreed on a plan to build the largest ships in the world—including the *Titanic*, that infamous vessel of modernity that would steam across the Atlantic in April 1912 to its ignominious end.

From Captain Edward Smith's point of view, the "unsinkable" ship striking an iceberg, and that collision causing its loss and the deaths of more than 1,500 passengers, was very much unexpected. Warnings of icebergs sighted in his route had reached Smith, but he only incrementally adjusted the ship's course further south and continued at full speed, warnings or no. His decisions were consistent with then-current practice which held that even a direct iceberg collision would not likely damage this kind of ship, much less sink it. Thus, to not just the *Titanic*'s captain but also the crew and passengers, and most of the world at that time, the *Titanic* was unlucky in encountering that lone iceberg floating in its path, and even more unfortunate that the encounter ripped a long enough gash in its side to send the ship plunging to the icy depths.

However, there is much more to this story. The *Carpathia*, the ship that upon hearing *Titanic*'s distress signals steamed fifty-eight miles to its rescue, covered the distance at maximum speed, but just as it got close enough to witness green flares fired from lifeboats, its progress slowed to a crawl. The ship was forced to navigate a vast ice field extending for miles across the area near where the *Titanic* was then sinking.[31] In the words of the *Carpathia*'s captain, "about two or three miles from the position of the *Titanic*'s wreckage we saw a huge ice field extending as far as we could see, N.W. to S.E." A *Titanic* crew survivor who piloted one of the lifeboats reported similarly: "In the morning, when it turned daybreak, we could see icebergs everywhere; also, a field of ice about 20 to 30 miles long. . . . The icebergs was [*sic*] up on every point of the compass, almost."[32] As it turns out, unusually warm conditions and wind patterns months before in early 1912 had conspired to place many more icebergs than normal, and in locations further south than usual, all the way to the Grand Banks off Newfoundland where the *Titanic* cruised silently that April night.[33]

In truth, then, the *Titanic* did not strike an errant, lonely iceberg floating in the North Atlantic—it struck the first one it encountered in an area littered with them for miles and miles, a veritable Iceberg Alley, as these instances of ice expulsions drifting south from Greenland were known. Steaming at full speed that night, the *Titanic* was not at risk of striking an iceberg, as most of us have understood—it was at risk of striking tens or even hundreds of them, all lying silently in wait in its path.

Recently, new scientific investigation has shone an intriguing light upon the larger context of the situation the *Titanic* faced that spring. A 2016 study[34] looking into the early manifestations of climate change concludes that warming of the oceans has been occurring for a far longer time than previously believed—a finding, incidentally, with which Rachel Carson would have agreed, confirmatory as it is of her own observations of Arctic warming in the 1940s that puzzled her so, as discussed in chapter 9. This study finds that "sustained, significant warming" began in the oceans not in recent decades, not even over the last century, but all the way back to the early nineteenth century. The study finds evidence of warming beginning in the 1830s, a time frame that puts the observations just decades after the burst of emissions marking the start of the Industrial Revolution. The conclusion is also consistent with the fact that, due to seawater covering so much of the earth's surface and being of a dark color that reflects little light back to space, the earth's oceans bear the brunt of global warming, absorbing more than 90 percent of the human-caused planetary warming. Notably, this study also found particularly sustained warming in the Northern Hemisphere and the Arctic region, including Greenland, which spawned *Titanic*'s iceberg.[35]

Did climate change sink the *Titanic*? Certainly not. However, just as the quickening pace of climate catastrophes we see all around us increasingly forces us to reconsider our future, so will we in time come to reassess our history in ways that piece together more completely the story of how exactly we have come to arrive at the situation we are in. This would mean fewer history books full of derring-do, conquest, and high ideals, and more with tales of carbon intensity, the climate-industrial complex, and iterating geochemistry. Just as we may come to view the lessons of Indianola and of Galveston and its seawalls differently through a new historical lens, we will perhaps come to understand all our modern human history as the story of how one species altered the planet's fundamental systems within a fraction of a geologic-time nanosecond. And in this regard, it is reasonable to recognize that unusually warm springs like that of 1912, which set in motion the sequence of events that placed an Iceberg Alley unusually early that year right in the *Titanic*'s route, became more likely due to the near century of incremental emissions-driven warming that the North Atlantic had already experienced before the idea of the *Titanic* was even conceived.

As the years go by now, each more clearly revealing an accelerating warming across our planet, the central paradox observed by Naomi Oreskes' and Erik Conway's fictional twenty-fourth-century historian in their book *The Collapse of Western Civilization* will haunt us: "Western civilization had the technological know-how and capability to effect an orderly transition to renewable energy, yet the available technologies were not implemented in time." How puzzling it indeed is that we understand so clearly not just what causes climate change but also the myriad and grave problems it will create, the threats these problems pose to our species, most particularly to our most vulnerable populations, and, beyond that, to the very web of life that sustains us—all while knowing full well the steps needed to stop it.

In 2023, the mirror speaks back to us—are we but twenty-first-century *Titanic* passengers, sailing upon our ship of modern life through an impossibly beautiful night sea that is the amazing Garden of Eden that the natural world has been for us during our time? Are we those travelers dancing gaily upon our imperturbable voyage speeding ahead, tacking gently with each passing moment all the more certainly toward a deadly adversity rising on the horizon? Are we that ship's captain, blithely noting the warnings of danger ahead but taking only the least inconvenient corrective actions while barreling full steam ahead? And as terrible events, so amply foretold, unfold around us, will we fail to prioritize the most vulnerable among us for the safety available? All the while knowing how this story ends?

To the extent we fail to meet the challenge before us, climate change becomes a zero-sum negotiation of quality of life between those of us alive today and those in future generations. Billions of us are seated on one side of the table, and only our consciences on the other side.

The greater the crisis becomes, the greater the opportunity before us to change the world, much as generations before us achieved the epic accomplishments that changed their worlds for the better in all the ways we so appreciate today. The greatest struggles for progress in our modern history are astonishingly impressive—the fight to end slavery and then Jim Crow in the United States, the fight for women's suffrage all around the world, the fight to beat global fascism in World War II, the fight to end colonial rule in India and to reverse apartheid in South Africa. The struggle to engage effectively on the problem of climate change is not so different from these that we cannot take all the inspiration we need to go forward with confidence enough to meet this challenge.

ACKNOWLEDGMENTS

H aving never written a book before, I needed a lot of help. First and foremost, this volume would not be in your hands today without S. Kirk Walsh, a talented author, editor, and friend. Kirk saw the book that was hiding in the manuscript I had written and then helped me find it. I am extremely grateful for her assistance. I also am very grateful to Casey Kittrell, who provided the crucial encouragement to me to embark upon this project and believed in my ability to pull it off.

A great thanks to Travis Snyder and the rest of the team at Texas Tech University Press. Travis guided me through editing and the remainder of the publishing process wisely and patiently. I am also greatly indebted to Katharine Hayhoe at Texas Tech for her involvement. She saw merit even in an early version of the manuscript, and I am so honored and pleased to have her foreword.

This book was mostly written from 2016 to 2019 and required a good deal of research and editing over that period. Heartfelt thanks to Taylor Lovely, Tim McCollough, and Regina Buono for their work editing, proofing, and researching—although any and all errors remain mine, of course.

Finally, my deepest thanks to my family, who have steadfastly believed this book could be written and one day published. My mother always believes her children can do what they set their minds to. My wife and kids put up with this project for so long. Many days a question would be posed—"Dad, can we go for a bike ride?" or "Andy, are you ready to leave yet?"—with the answer so often, "No, I'm sorry, I'm working on my book . . ." I could not have undertaken this project nor finished it without the love and encouragement of my wife Rachel, who also wins the prize for reading the most manuscript revisions.

NOTES

FOREWORD

1. The United States is responsible for 26 percent of global carbon dioxide emissions since the dawn of the industrial era. Its contribution is double that of any other country. China is in second place, with nearly 13 percent of global emissions, and Russia is third, with 7 percent.

CHAPTER 1

1. There are numerous compelling accounts of the 1900 Storm, but none are better than Erik Larson's *Isaac's Storm, A Man, A Time and the Deadliest Hurricane in History*, 2000, Vintage.
2. "Portrait of a Legend: The Great Storm of 1900: St. Mary's Orphan Asylum," *Houston Family Magazine*, December 31, 2012, https://www.houstonfamilymagazine.com/features/portrait-of-a-legend-the-great-storm-of-1900-st-marys-orphan-asylum/.
3. Isaac M. Cline, Special Report on The Galveston Hurricane of September 8, 1900, *Weather Bureau* (Report), September 23, 1900, Galveston, Texas, 374, accessed June 22, 2019, https://www.aoml.noaa.gov/general/lib/lib1/nhclib/mwreviews/1900.pdf.
4. "Galveston: The Mother of All U.S. Natural Disasters," *PBS News Hour*, September 28, 2011, https://www.pbs.org/newshour/world/galveston-the-mother-of-all-us-natural-disasters.

5. "Galveston, the City Reclaimed," *Pearson's Magazine* 13, no. 3 (March 1905).
6. "Galveston Seawall and Grade Raising Project," American Society of Civil Engineers website, accessed April 20, 2018, https://www.asce.org/project/galveston-seawall-and-grade-raising-project/.
7. "Galveston, the City Reclaimed."
8. "Weathering the Storm: The Galveston Seawall and Grade Raising," *Civil Engineering* 77, no. 4 (April 2007): 32–33, https://ascelibrary.org/doi/pdf/10.1061/ciegag.0000797.
9. "Galveston's Bulwark Against the Sea," Army Corps of Engineers report, revised October 1981, http://www.swg.usace.army.mil/Portals/26/docs/PAO/GalvestonBulwarkAgainsttheSea.pdf.
10. "Galveston's Bulwark Against the Sea"; "Galveston Seawall and Grade Raising Project"; and "Weathering the Storm."
11. "Galveston, the City Reclaimed."
12. Vidor's account of his experience surviving the storm was published as "Southern Storm" in *Esquire*, May 1, 1935.
13. "How Much Power Is a Gigawatt," U.S. Department of Energy (DOE), accessed June 26, 2019, https://www.energy.gov/eere/articles/how-much-power-1-gigawatt.
14. "2040 RTP Demographics," Houston-Galveston Area Council, accessed April 18, 2018, http://www.h-gac.com/taq/plan/2040/demographics.aspx.
15. Ibid.
16. "Energy and the Environment Explained: Where Greenhouse Gases Come From," US Energy Information Administration, accessed April 28, 2021, https://www.eia.gov/energyexplained/energy-and-the-environment/where-greenhouse-gases-come-from.php.
17. "How the World Passed a Carbon Threshold and Why It Matters," Yale Environment 360, Yale School of Forestry and Environmental Studies, January 26, 2017, https://e360.yale.edu/features/how-the-world-passed-a-carbon-threshold-400ppm-and-why-it-matters.
18. "CO_2 Levels Continue to Increase at Record Rate," Yale Environment 360, March 14, 2017, https://e360.yale.edu/digest/co2-levels-continue-to-increase-at-record-rate.

CHAPTER 2

1. Subsequent to the publication of the first edition of this book, the IPCC issued the Sixth Assessment Report, which can be found at https://www.ipcc.ch/assessment-report/ar6/.

2. The IPCC report classes its findings on the kinds of consequences we can expect into different "Key Risks," each in turn associated with one or more of a handful of broader categories that it presents in its discussion of risks across sectors and regions. Risk is forecast for each of the broader categories along a scale of expected temperature increases: Increases of 1 or 2 degrees Celsius heighten the risk level for each one at different rates, but every one in this range reaches "Moderate" or "High" risk levels, and increases beyond 2 degrees escalate to "Very High." Since these are projections about the future, the authors are careful to explicitly state their confidence level regarding the projections. For the Key Risks, they note "high confidence."

3. Brad Plumer, "Two degrees: The world set a simple goal for climate change. We're likely to miss it." *Vox*, April 22, 2014, https://www.vox.com/2014/4/22/5551004/two-degrees. The article attributes the observation regarding the connection between 2 degrees Celsius and human civilization to German physicist and climatologist Hans Joachim Schellnhuber, one of the scientists on the German advisory panel who helped devise the 2°C limit. Schellnhuber recounted, "We said that, at the very least, it would be better not to depart from the conditions under which our species developed. Otherwise, we'd be pushing the whole climate system outside the range we've adapted to."

4. "Paris Agreement," Section 1(a), United Nations, 2015, http://unfccc.int/files/essential_background/convention/application/pdf/english_paris_agreement.pdf.

5. "Climate Science: Special Report," Fourth National Climate Assessment, 2017, U.S. Global Change Research Program (USGCRP), 396. Report may be viewed in full at https://science2017.globalchange.gov/downloads/CSSR2017_FullReport.pdf.

6. Ibid., 396.

7. Ibid., 397.

8. Ibid., 397.

9. Jeff Tollefson, "COVID Curbed Carbon Emissions in 2020—But Not By Much," *Nature*, January 15, 2021, accessed March 21, 2021, https://www.nature.com/articles/d41586-021-00090-3; Jeff Tollefson, "How the Coronavirus Pandemic Slashed Carbon Emissions—In Five Graphs," *Nature*, May 20, 2020, accessed March 21, 2021, https://www.nature.com/articles/d41586-020-01497-0; Lauri Myllyvirta, "Analysis: China's CO_2 Emissions Surged 4% in Second Half of 2020," Carbon Brief, March 1, 2021, accessed March 27, 202, https://www.carbonbrief.org/analysis-chinas-co2-emissions-surged-4-in-second-half-of-2020

10. "Climate Science: Special Report," USGCRP, 397.

11. Cindy L. Bruyère, "Impact of Climate Change on Gulf of Mexico Hurricanes," National Center for Atmospheric Research (NCAR), accessed June 25, 2019, https://www.c3we.ucar.edu/impact-climate-change-gulf-mexico-hurricanes

12. Climate Change Projections for the City of Austin, Draft Report for 2014, Atmos Research and Consulting, accessed June 30, 2019, https://austintexas.gov/sites/default/files/files/Sustainability/atmos_research.pdf.

13. "Coronavirus, Climate Change, and the Environment, A Conversation on COVID-19 with Dr. Aaron Bernstein, Director of Harvard Chan C-CHANGE," accessed November 15, 2020, https://www.hsph.harvard.edu/c-change/subtopics/coronavirus-and-climate-change/

14. R. B. Jackson et al., "Persistent Fossil Fuel Growth Threatens the Paris Agreement and Planetary Health," Environ. Res. Lett. 14 121001, accessed March 21, 2021, https://iopscience.iop.org/article/10.1088/1748-9326/ab57b3.

15. Chelsea Harvey, "Global Carbon Emissions Are Rising Again after 3 Flat Years," *Scientific American*, November 13, 2017, https://www.scientificamerican.com/article/global-carbon-emissions-are-rising-again-after-3-flat-years/.

16. "Air Quality Index (AQI) Basics," AirNow, accessed June 25, 2019, https://airnow.gov/index.cfm?action=aqibasics.aqi.

17. K. Vohra et al., "Global Mortality from Outdoor Fine Particle Pollution Generated by Fossil Fuel Combustion: Results From GEOS-Chem," *Environmental Research* 195, April 2021,

110754, accessed March 21, 2021, https://doi.org/10.1016/j. envres.2021.110754

18. Timothy Cama, "India Air Took 6 Hours off Obama's Life," *The Hill*, January 27, 2015, accessed June 25, 2019, https://thehill.com/policy/energy-environment/230829-indi a-air-took-6-hours-off-obamas-life.

19. Dennis Silverman, "Imported Sources of California's Electric Power in 2014, and its CO_2 Pollution," *Energy Blog*, November 25, 2015, accessed June 25, 2019, https://sites.uci.edu/ energyobserver/2015/11/25/sources-of-californias-electri c-power-in-2014-and-its-co2-pollution/.

20. Douglas J. Arent et al., "Key Economic Sectors and Services," *Climate Change 2014: Impacts, Adaptation, and Vulnerability—Part A: Global and Sectoral Aspects* (New York: Cambridge University Press, 2014), 665.

21. Peter Brannen, "Mainstreaming Agrobiodiversity in Sustainable Food Systems," Bioversity International, 2016, https://www.bioversityinternational.org/fileadmin/user_upload/Mainstreaming_Summary_2017.pdf.

22. Yinon M. Bar-On, Rob Phillips, and Ron Milo, "The Biomass Distribution on Earth," Proceedings of the National Academy of Sciences 115, no. 25 (June 19, 2018), accessed March 21, 2021, https://doi.org/10.1073/pnas.1711842115

23. R. K. Pachauri and L. A. Meyer, eds., "Climate Change 2014, Synthesis Report, Summary for Policymakers, *Climate Change 2014: Synthesis Report. Contribution of Working Groups I, II and III to the Fifth Assessment Report of the Intergovernmental Panel on Climate Change*, (Geneva, Switzerland: Intergovernmental Panel on Climate Change, 2014), 14, accessed November 15, 2020, https://www.ipcc. ch/site/assets/uploads/2018/02/AR5_SYR_FINAL_SPM.pdf

24. Subsequent to the publication of the first edition of this book, the IPCC issued the Sixth Assessment Report, which can be found at https://www.ipcc.ch/assessment-report/ar6/

25. Paris Agreement Tracker, Climate Action Tracker, December 2020, accessed March 21, 2021, https://climateactiontracker.org/global/ cat-thermometer/

26. "The available scenarios [for 1.5°C and 2.0°C] show rapidly declin-

ing emissions after 2030, with global CO_2 emissions from energy-
and industry-related sources reaching net-zero levels between
2060 and 2080." In Joeri Rogelj et al., "Paris Agreement Climate
Proposals Need a boost to Keep Warming Well Below 2°C," *Nature*
534 (June 29, 2016), http://www.readcube.com/articles/10.1038/
nature18307?no_publisher_access=1&r3_referer=nature&refer-
rer_host=www.nature.com.

CHAPTER 3

1. "Fort Stockton, Texas," Best Places, accessed June 9, 2018, https://
   www.bestplaces.net/climate/city/texas/fort_stockton.
2. "Fort Stockton," The Handbook of Texas Online, accessed June 9,
   2018, https://tshaonline.org/handbook/online/articles/qbf46.
3. "Loss of Comanche Springs Shadows Balmorhea Issue," *Houston
   Chronicle*, October 15, 2016, https://www.houstonchronicle.
   com/business/energy/article/Loss-of-Comanche-Springs-shad-
   ows-Balmorhea-issue-9973094.php.
4. Daniel Yergin, *The Prize: The Epic Quest for Oil, Money and Power*
   (New York: Free Press, 1991), 223.
5. "Yates Oil Field—Oil Patch History," Landman Blog, accessed
   August 2, 2018, http://www.landmanblog.com/yates-oil-field/.
6. "History of the Yates Oilfield Near the Town of Iraan, in West
   Texas," Energy Industry Photos, accessed August 2, 2018, http://
   www.energyindustryphotos.com/history_of_the_yates_oil_field_n.
   htm.
7. Ibid.
8. "Yates Oilfield, West Texas," NASA Earth Observatory, accessed
   May 6, 2018, https://earthobservatory.nasa.gov/IOTD/view.
   php?id=6776.
9. Julia Cauble Smith, "Yates Oilfield," TSHA Online, accessed August
   2, 2018, https://tshaonline.org/handbook/online/articles/doy01.
10. Brett Clanton, "Houston Company Nudges Oil Out of Geriatric
    Field," *Houston Chronicle*, June 27, 2011, https://www.chron.com/
    business/energy/article/Houston-company-nudges-oil-out-of-geria
    tric-field-2081041.php.
11. For more on this topic, see Kate Galbraith and Asher Price, *The
    Great Texas Wind Rush: How George Bush, Ann Richards, and a*

Bunch of Tinkerers Helped the Oil and Gas State Win the Race to Wind Power (Austin: University of Texas Press, 2013).

12. "First Solar Completes Barilla Solar Project," First Solar, September 4, 2014, http://investor.firstsolar.com/news-releases/news-release-details/first-solar-completes-barilla-solar-project.

13. Calculated from "Monthly Energy Review," March 2018, Table 7.2A, p. 111, U.S. Energy Information Administration, https://www.eia.gov/totalenergy/data/monthly/archive/00351803.pdf.

14. World Energy Balances: Overview (2018 edition), International Energy Agency, 4.

15. A great primer on utility regulation's past and present is *Electricity Regulation in the US: A Guide*, 2011, by Jim Lazar with the Regulatory Assistance Project (RAP), free online at https://www.raponline.org/wp-content/uploads/2016/05/rap-lazar-electricityregulationintheus-guide-2011-03.pdf. The guide summarizes the early history of utility regulation as follows: "Initially, electric and gas utilities competed with traditional fuels (e.g., peat, coal, and biomass, which were locally and competitively supplied), and were allowed to operate without regulation. If they could attract business, at whatever prices they charged, they were allowed to do so. Cities did impose 'franchise' terms on them, charging fees and establishing rules allowing them to run their wires and pipes over and under city streets. Around 1900, roughly twenty years after Thomas Edison established the first centralized electric utility in New York, the first state regulation of electric utilities emerged" (7–8).

16. "Starting in the 1970s, higher fuel prices, environmental concerns, technological innovations and a desire for more economic efficiency led to the rethinking of this vertically integrated, regional monopoly model." "The History and Evolution of the U.S. Electricity Industry," UT Austin Energy Institute, July 2016, p. 1, http://sites.utexas.edu/energyinstitute/files/2016/09/UTAustin_FCe_History_2016.pdf. This source provides an excellent overview of the history of US utility regulation.

17. Ibid., 11.

18. For background on the phenomenon of fossil fuel merchant plants, see, for instance, Carrie La Seur, "How Merchant Coal Is

Changing the Face of America," *Grist*, August 24, 2006, https://grist.org/article/laseur/, and Brian K. Schimmoller, "The Merchants Are Coming," Power Engineering, August 1, 1998, https://www.power-eng.com/articles/print/volume-102/issue-8/features/the-merchants-are-coming.html.

19. Wind energy projects have also not been financeable on a merchant basis, although for different reasons. While wind energy has been priced more cheaply than prevailing wholesale electricity for several years in many US markets, wind still requires PPAs mainly because wind financing involves significant volumes of tax credits. Tax investors participate in wind project financing to monetize these credits, and these investors' risk tolerances are so low that they have refused to participate. Wind tax credits are expected to expire in the next few years, however, perhaps providing a basis for merchant wind projects to appear by the early 2020s.

20. Gavin Weightman, *Children of Light: How Electricity Changed Britain Forever* (London: Atlantic Books, 2011), 67.

CHAPTER 4

1. There has been a lively ongoing debate on the utility death spiral idea. For additional reading, see Peter Kind, *Disruptive Challenges: Financial Implications and Strategic Responses to a Changing Retail Electric Business* (Edison Electric Institute, 2013), https://books.google.com/books/about/Disruptive_Challenges.html?id=sCYsmwEACAAJ, and William Pentland, "Disruption Derailed," *Forbes*, November 25, 2015, https://www.forbes.com/sites/williampentland/2015/11/25/disruption-derailed-the-utility-death-spiral-myth/#2bb51cd65d1f. More recently, see Russell Ray, "Are Electric Utilities in a Death Spiral?" *Energize Weekly*, September 26, 2018, https://www.euci.com/are-electric-utilities-in-a-death-spiral/, noting that "A whopping 71 percent of utilities said they believe the death spiral is a real and possible outcome if the industry fails to implement alternative energy solutions and/or regulations fail to recognize flexibility."

2. Klaus Bader, "Corporate PPAs," Norton Rose Fulbright, March 9, 2017, http://www.nortonrosefulbright.com/files/9_corporate-ppas-147067.pdf.

3. See, for example, Frank Swigonski, "Corporate Procurement of Renewable Energy Gets Creative," Advanced Energy Perspective's blog, April 18, 2017, https://blog.aee.net/corporate-procuremen t-of-renewable-energy-gets-creative.

4. Diane Cardwell, "Apple Becomes a Green Energy Supplier, With Itself as Customer," *New York Times*, August 23, 2016, https:// www.nytimes.com/2016/08/24/business/energy-environment/ as-energy-use-rises-corporations-turn-to-their-own-gre en-utility-sources.html.

5. "Pattern and Walmart Dedicate 200 MW Logan's Gap Wind Farm," North American Windpower, accessed June 25, 2019, https://nawindpower.com/pattern-and-walmart-dedicat e-200-mw-logans-gap-wind-farm.

6. Amy Brown, "Walmart Scales Up Wind Power Purchases as it Eyes 100 Percent Renewable Energy," Triple Pundit, accessed June 25, 2019, https://www.triplepundit.com/story/2018/walmart-scale s-wind-power-purchases-it-eyes-100-percent-renewable-en-ergy/10456.

7. "Natural Gas-Fired Generating Capacity Likely to Increase Over Next Two Years," U.S. Energy Information Administration web-site, January 30, 2017, https://www.eia.gov/todayinenergy/detail. php?id=29732.

8. IEA, Natural Gas-Fired Power, accessed April 28, 2021, https:// www.iea.org/reports/natural-gas-fired-power..

9. The idea that economic forces are a safer long-term bet for reducing emissions than are politics and policy does have its Achilles heel. If future innovations in fossil fuel power generation allow these plants to compete on price once again with renewable energy, or if economic nationalism results in trade impediments on solar and wind equipment from the cheapest countries—a factor very much at play in the world today—then emissions would be positioned to increase substantially.

CHAPTER 5

1. Philip Jaekl, "Why People Believe Low-Frequency Sound Is Danger-ous," *The Atlantic*, accessed June 25, 2019, https://www.theatlantic. com/science/archive/2017/06/wind-turbine-syndrome/530694/.

2. Ibid.

3. Loren D. Knopper and Christopher A. Ollson. "Health Effects and Wind Turbines: A Review of the Literature," *Environmental Health: A Global Access Science Source*, vol. 10, September 14, 2011, 78, doi:10.1186/1476-069X-10-78, https://www.ncbi.nlm.nih.gov/pmc/articles/PMC3179699/

4. Stephen Moore and Julian L. Simon, "The Greatest Century That Ever Was: 25 Miraculous Trends of the Past 100 Years," *Policy Analysis* no. 364, Cato Institute (December 15, 1999): 1–32. "Although generation of electrical power was possible by the late nineteenth century, electricity started to become widely available in homes and factories only in the early decades of this century" (4).

5. Andy Bowman, "Bowman: Rick Perry Has Shown His True Colors, and They Aren't Green," *Houston Chronicle*, February 8, 2017, https://www.houstonchronicle.com/opinion/outlook/article/Bowman-Rick-Perry-has-shown-his-true-colors-and-10918648.php.

6. See Molly F. Sherlock and Jeffrey M. Stupak, "Energy Tax Incentives: Measuring Value Across Different Types of Energy Resources," Congressional Research Service Report for Congress, March 19, 2015; see also its antecedent report, Molly F. Sherlock, "Energy Tax Policy: Historical Perspectives on and Current Status of Energy Tax Expenditures," CRS Report for Congress, May 2, 2011.

7. Mona Hymel, "The United States' Experience with Energy-Based Tax Incentives: The Evidence Supporting Tax Incentives for Renewable Energy," *Loyola University Chicago Law Journal* 38, no. 1 (Fall 2006): 71–72.

8. "Updated Study Confirms Federal Energy Incentives Have Chiefly Benefited Oil, Natural Gas industries," PR Newswire, October 25, 2011, accessed April 28, 2021, https://www.prnewswire.com/news-releases/updated-study-confirms-federal-energy-incentives-have-chiefly-benefited-oil-natural-gas-industries-132521513.html.

9. Roger Bezdek, "US Energy Subsidies in Perspective" *Cornerstone, The Official Journal of the World Coal Industry* (Spring 2013): 31.

10. Nancy Pfund and Ben Healey, "What Would Jefferson Do? The Historical Role of Federal Subsidies in Shaping America's Energy Future," DBL Investors, September 2011, accessed April 28, 2021,

https://www.dblpartners.vc/wp-content/uploads/2012/09/
What-Would-Jefferson-Do-2.4.pdf?597435=&48d1ff=

11. Celeste Wanner, "New Analysis: Wind energy less than 3 percent of all federal energy incentives," American Wind Energy Association (AWEA), July 20, 2016.

12. Lukas Ross, "$135 Bullion Reasons Rex Tillerson Is Wrong," The Hill, January 17, 2017, accessed April 28, 2021, https://thehill.com/blogs/pundits-blog/energy-environment/314634-13 5-billion-reasons-rex-tillerson-is-wrong

13. Jason Burwen and Jane Flegal, "Unconventional Gas Exploration and Production," American Energy Innovation Council, March 2013, http://americanenergyinnovation.org/wp-content/uploads/2013/03/Case-Unconventional-Gas.pdf.

14. Ibid.

15. See Anthony Lopez, Billy Roberts, Donna Heimiller, Nate Blair, and Gian Porro, "U.S. Renewable Energy Technical Potentials: A GIS-Based Analysis," National Renewable Energy Laboratory, 2012, https://www.nrel.gov/docs/fy12osti/51946.pdf.

16. U.S. DOE, Energy Efficiency & Renewable Energy, *2015 Renewable Energy Data Book*, accessed March 24, 2021, https://www.nrel.gov/docs/fy17osti/66591.pdf

17. Some corporations do not disclose the capital costs of their projects for competitive reasons, and this appears to be the case with First Solar and Barilla. However, in one of the legal agreements between the Barilla project and a local government entity, First Solar described the project as having a capital cost of $75 million at a size of 50 megawatts, although the project actually constructed that is the subject of this book was only 18 megawatts in size. This implies that the cost of the 18 megawatt project was likely (18/50) X $75 million, or $27 million. See "Summary of Financial Impact of the Proposed Barilla Solar LLC Project on the Finances of the Fort Stockton Independent School District Under a Requested Chapter 313 Property Value Limitation," August 17, 2013, Final Report, and Exhibit 3 of the "Agreement for Limitation on Appraised Value of Property for School District Maintenance and Operations Taxes, by and between Fort Stockton Independent School District and Barilla Solar, LLC," December 16, 2013.

18. Richard A. Kessler, "We're Bullish on Texas: It's a Market Where We Want to Be a Big Player," Recharge: The Global Source for Renewable Energy News & Intelligence, September 9, 2015, http://www.rechargenews.com/solar/868861/were-bullish-on-texas-its-a-market-where-we-want-to-be-a-big-player.

19. "First Solar (FSLR) Q4 2016 Results—Earnings Call Transcript," Seeking Alpha, February 21, 2017, https://seekingalpha.com/article/4048085-first-solar-fslr-q4-2016-results-earnings-call-transcript#ampshare=https://seekingalpha.com/article/4048085-first-solar-fslr-q4-2016-results-earnings-call-transcript.

20. First Solar 2013 10K, p. 50 and footnote 11 on p. 52 table.

21. Marc Klezczewski, "Solar Continues to Lead Renewables Boom," *GCX Magazine*, June 27, 2014, http://gcxmag.com/2014/06/solar-continues-lead-renewables-boom/.

22. Ibid.

23. Jennifer Evans, "Rice Signs Landmark Energy Agreement with MP2 Energy," Rice University, January 29, 2015, http://news.rice.edu/2015/01/29/rice-signs-landmark-energy-agreement-with-mp2-energy/.

24. Jennifer Runyon, "New Deal Allows Rice University to 'Test' Solar Power Under Short-term Contract," Renewable Energy World, February 6, 2015, http://www.renewableenergyworld.com/articles/2015/02/new-deal-allows-rice-university-to-test-solar-power-under-short-term-contract.html.

25. "First Solar (FSLR) Q4 2016 Results—Earnings Call Transcript," accessed May 11, 2021, https://seekingalpha.com/article/4048085-first-solar-fslr-q4-2016-results-earnings-call-transcript.

26. Natural gas prices were $1.95 per million BTU in early 2012 and reached $6 in 2014. See Natural Gas: Henry Hub Natural Gas Spot Price, U.S. Energy Information Administration, https://www.eia.gov/dnav/ng/hist/rngwhhdm.htm.

27. "Crude Oil Prices: 70 Year Historical Chart," Macrotrends, accessed August 7, 2018, http://www.macrotrends.net/1369/crude-oil-price-history-chart.

28. Natural Gas: Henry Hub Natural Gas Spot Price.

29. ERCOT power prices by year are reported in "2016 State of the

Market Report for the ERCOT Electricity Markets," May 2017, iii; "2012 State of the Market Report for the ERCOT Wholesale Electricity Markets," June 2013, ii; and "2010 State of the Market Report for the ERCOT Electricity Markets," August 2011, v. All reports published by Potomac Economics Ltd: Independent Market Monitor for the ERCOT Wholesale Market.

30. "Panda Temple I Plant Files for Chapter 11," May 11, 2017, *Fort Worth Star-Telegram*, staff writer.

31. Deloitte, "A Brand New Merchant World," June 2020, 8, https://www2.deloitte.com/content/dam/Deloitte/dk/Documents/energy-resources/Downloads/Onshore_Renewables_Report_2020_FINAL.pdf

32. PR Newswire/Advanced Power, "Advanced Power achieves financial close of 140-MWdc cutlass Solar project in Fort Bend County, Texas," April 28, 2021.

33. Liam Stoker, "BP backs bifaciality, merchant revenue strategies to boost solar out to 2030", PVTech, September 16, 2020.

34. Christian Roselund, "Is the U.S. Solar Market Slipping Towards Merchant?" *PV Magazine*, June 18, 2019, accessed July 3, 2019, https://pv-magazine-usa.com/2019/06/18/is-the-u-s-solar-market-slipping-towards-merchant/.

35. Christian Roselund, "Duke Buys a 200 MW Merchant Solar Project in Texas," *PV Magazine*, July 23, 2019, accessed November 28, 2020, https://pv-magazine-usa.com/2019/07/23/duke-buys-a-200-mw-merchant-solar-project-in-texas/.

36. Marco Dorothal, "Top 15 Solar Merchant Projects," SolarPlaza, accessed September 6, 2018, https://www.solarplaza.com/channels/top-10s/11818/top-15-merchant-solar-plants/.

37. Jonathan Gifford, "Risen Breaks Ground on 'Merchant' 121 MW Yarranlea Project," *PV Magazine*, May 24, 2018; Lynne Grierson, "Sod Turned on Northam Solar Farm Project," *Hills Gazette* (Perth) March 20, 2018; Giles Parkinson, "Kidston Solar Farm Lays Out Case for 'Going Merchant,'" Renew Economy, accessed September 6, 2018, https://reneweconomy.com.au/kidston-solar-farm-lays-case-going-merchant-84317/; Sophie Vorrath, "Chinchilla Solar Farm Secures Debt Funding to 'Go Merchant,'" Renew Economy, accessed September 6, 2018, https://reneweconomy.com.au/

chinchilla-solar-farm-secures-debt-funding-go-merchant-27047/.

38. "Danish Merchant Solar Set for Take-Off—At Former Airfield," Powerlinks News, December 13, 2019, accessed November 28, 2020, https://powerlinks.news/denmark/news/danish-merchan t-solar-set-for-take-off-at-former-airfield

39. Jose Rojo, "Post-COVID Merchant Solar: The Financier View of Banco Sabadell," PV-Tech, May 2020, accessed November 28, 2020, https://solar-media.s3.amazonaws.com/assets/Pubs/PVTP%2023/ Post-COVID%20merchant%20solar%20-%20The%20financier%20 view%20of%20Banco%20Sabadell.pdf

40. Author's direct communications with wind project owners / operators.

41. Tino Andresen, "Offshore Wind Farms Offer Subsidy-Free Power for First Time," Bloomberg, April 13, 2017, https://www. bloomberg.com/news/articles/2017-04-13/germany-gets-bids-fo r-first-subsidy-free-offshore-wind-farms.

42. Pilar Sanchez Molina, "Renovalia Nails Financing for Spain's First Merchant PV Project," *PV Magazine*, August 5, 2019, accessed November 28, 2020, https://www.pv-magazine.com/2019/08/05/ renovalia-nails-financing-for-spains-first-merchant-pv-project/.

43. "Annual Energy Outlook 2019," U.S. Energy Information Administration, January 24, 2019, Slide 95, accessed May 26, 2019, https:// www.eia.gov/outlooks/aeo/pdf/aeo2019.pdf.

44. Ibid.

CHAPTER 6

1. Roger Mudd, "White House / Water Heater," Television Vanderbilt News Archive, *NBC Evening News*, August 22, 1986, https:// tvnews.vanderbilt.edu/broadcasts/551774

2. K. Dunbar and J. Fugelsang, "Causal Thinking in Science: How Scientists and Students Interpret the Unexpected," in *Scientific and Technological Thinking*, ed. M. E. Gorman, R. D. Tweney, D. Gooding, and A. Kincannon (Mahwah, NJ: Lawrence Erlbaum Associates, 2005), 57–79.

3. Ibid.

4. I am engaging in some conjecture in asserting that Edmond Becquerel's discovery occurred in the basement. It is documented that he

worked in his father's lab located in Paris's Jardin des Plantes (now headquarters of the *Muséum nationale d'histoire naturelle*), which has large basement spaces, and that he completed important steps of the experiment in a "darkroom" that was capable of providing absolute darkness, so it seems a reasonable assumption.

5. For a detailed description of Becquerel's famous experiment and discussion of his prowess at obtaining exact measurements using nineteenth-century scientific instruments, see Jérôme Fatet, "Recreating Edmond Becquerel's electrochemical actinometer," *Archives Des Sciences Journal* 5 (September 2005).

6. A book will someday be written about the teenaged Becquerel and the details of his remarkable discovery, but until then we can only guess what he was doing experimenting with electrolytic cells in his father's laboratory at that age and how and why he came up with the experiment that led to the discovery in the first place. History must wait to see whether Becquerel's gift to humanity, discovering the photovoltaic effect, will ultimately be larger than that of his son, Henri, who, along with Marie Curie and others, discovered radioactivity—but it is worth pointing out that the younger received the Nobel Prize and the older did not.

7. John Perlin, *From Space to Earth: The Story of Solar Electricity* (Cambridge, MA: Harvard University Press, 2002), 15–17. Perlin's history of solar energy is a lively, authoritative chronology of solar technologies.

8. Ibid.

9. Ibid, 17.

10. Geoffrey Jones and Loubna Bouamane, "'Power from Sunshine': A Business History of Solar Energy," Harvard Business School Working Paper, May 25, 2012, https://hbswk.hbs.edu/item/power-from-sunshine-a-business-history-of-solar-energy.

11. "The Silicon Engine: Timeline," Computer History Museum, accessed April 10, 2017, http://www.computerhistory.org/siliconengine/timeline/.

12. "The Silicon Engine: 1874—Semiconductor Point-Contact Rectifier Effect Is Discovered," Computer History Museum, accessed April 10, 2017, http://www.computerhistory.org/siliconengine/semiconductor-point-contact-rectifier-effect-is-discovered/.

13. Jon Gertner, "True Innovation," *New York Times*, February 25, 2012.

14. The extraordinary role of Bell Labs in creating so much of the core technology of our world today is recounted in Jon Gertner's excellent book *The Idea Factory: Bell Labs and the Great Age of American Innovation* (New York: Penguin Books, 2012).

15. "Russell Ohl: Biography," Engineering and Technology History Wiki, last modified February 25, 2016, http://ethw.org/Russell_Ohl.

16. Jones and Bouamane, "'Power from Sunshine,'" 13, and Perlin, *From Space to Earth*, 28.

17. This observation is drawn from Perlin, *From Space to Earth*, 25–29.

18. Jones and Bouamane, "'Power from Sunshine,'" 13–14.

19. Perlin, *From Space to Earth*, 27–31.

20. Perlin, John. "The Invention of the Solar Cell," *Popular Science*, April 22, 2014.

21. Perlin, *From Space to Earth*, 27–31.

22. Sheila Bailey, Ryne Raffaelle, and Keith Emery, "Space and Terrestrial Photovoltaics—Synergy and Diversity," conference paper, 17th Space Photovoltaic Research and Technology Conference, October 2002.

23. Kat Eschner, "The World's First Solar-Powered Satellite is Still Up There After More Than 60 Years," March 17, 2017, Smithsonianmag.com.

24. The important role of the space program in advancing solar technologies during this period is detailed by Perlin in *From Space to Earth*, 41–47, and by Bailey, Raffaelle, and Emery in "Space and Terrestrial Photovoltaics."

25. Perlin, *From Space to Earth*: $286 in 1955 (p. 36); $100 in 1971 (p. 50); $11 in 1980 and $7.00 in 1985 (p. 118); $5.30 in the 1990s (p. 153). BNEF chart cited in Zachary Shahan, "13 Charts on Solar Panel Cost & Growth Trends," CleanTechnica, September 4, 2014, https://cleantechnica.com/2014/09/04/solar-panel-cost-trends-10-charts/

26. Tim McMahon, "Historical Crude Oil Prices (Table)," InflationData.com, August 27, 2017, https://inflationdata.com/Inflation/Inflation_Rate/Historical_Oil_Prices_Table.asp.

27. Eliza Griswold, "How 'Silent Spring' Ignited the Environmental Movement," *New York Times*, September 21, 2012, https://www.nytimes.com/2012/09/23/magazine/how-silent-spring-ignite d-the-environmental-movement.html.

28. "History of Solar Power," Institute for Energy Research, February 18, 2016, https://instituteforenergyresearch.org/analysis/history-of-solar-power/.

29. Jones and Bouamane, "'Power from Sunshine,'" 22–25.

30. See, for instance, Bailey, Raffaelle, and Emery, "Space and Terrestrial Photovoltaics."

31. Ibid., 28.

32. James Bates, "Sale of ARCO Unit Casts Shadow on Future of Solar Energy Ventures," *Los Angeles Times*, March 7, 1989, http://articles.latimes.com/1989-03-07/business/fi-233_1_solar-energy.

33. Tom Cheyney, "A Tribute to Solar PV Pioneer Bill Yerkes," Greentech Media, February 17, 2014, https://www.greentechmedia.com/articles/read/a-tribute-to-solar-pv-pioneer-bill-yerkes#gs.nhdG5rs.

34. "Plans to Build and Operate the World's Largest Solar . . . ," UPI, April 1, 1982, https://www.upi.com/Archives/1982/04/01/Plans-t o-build-and-operate-the-worlds-largrest-solar/6431386485200/.

35. Philip Wolfe, *Solar Photovoltaic Projects in the Mainstream Power Market* (New York: Routledge, 2010), 24.

36. McMahon, "Historical Crude Oil Prices (Table)," InflationData.com.

37. Jones and Bouamane, "'Power from Sunshine,'" 22–25.

CHAPTER 7

1. David Roberts, "The Falling Costs of U.S. Solar Power, in 7 Charts," Vox, August 24, 2016, https://www.vox.com/2016/8/24/12620920/us-solar-power-costs-falling, and Molly Cox, "Key 2020 US Solar PV Cost Trends and a Look Ahead," Green Tech Media, December 17, 2020.

2. Caroline Delbert, "This Solar Panel Just Set a World Record for Efficiency," *Popular Mechanics*, April 20, 2020.

3. "Solar Industry Research Data," SEIA, June 29, 2019, https://www.seia.org/solar-industry-research-data.

4. Shahan, "13 Charts on Solar Panel Cost & Growth Trends."

5. Snapshot of Global PV Markets – 2020, International Energy Agency (IEA) Technology Collaboration Programme, April 2020, https://iea-pvps.org/wp-content/uploads/2020/04/IEA_PVPS_Snapshot_2020.pdf

6. IEA (2020), Renewables 2020, IEA, Paris, https://www.iea.org/reports/renewables-2020

7. "Chernobyl Accident and Its Consequences," NEI Fact Sheet, March 2015, https://www.nei.org/resources/fact-sheets.

8. Paul Hockenos, "How Germany Learned to Hate Nuclear Power," chinadialogue, October 23, 2012, https://www.chinadialogue.net/article/show/single/en/5232-How-Germany-learned-to-hate-nuclear-power.

9. Paul Hockenos, "Energiewende—The First Four Decades," Clean Energy Wire, June 22, 2015, https://www.cleanenergywire.org/dossiers/history-energiewende.

10. Ibid.

11. Ibid.

12. Louise Branson, "Soviets Putting Moore Missile in East Germany," UPI, May 14, 1984, https://www.upi.com/Archives/1984/05/14/Soviets-putting-moore-missile-in-East-Germany/2417453355200/, and Stephen Evans, "A Soviet Missile Base in Germany That Spy Planes Never Saw," BBC News, October 26, 2012, http://www.bbc.com/news/magazine-20079147.

13. Hockenos, "Energiewende."

14. Craig Morris, "'Efficiency Lacks a Loud Lobby': An Interview with Florentin Krause," Energy Transition: The Global Energiewende (blog), April 17, 2013, https://energytransition.org/2013/04/an-interview-with-florentin-krause-2/. Krause, one of the authors of the book that coined the phrase "Energiewende," explained that "Germany is fortunate in that the grassroots environmental movement of the 1970s and 1980s was able to articulate itself in the electoral system in the form of the Green Party."

15. Jones and Bouamane, "Power from Sunshine," 49.

16. Rainer Quitzow, "Dynamics of a Policy Driven Market: The Co-Evolution of Technological Innovation Systems for Solar Photovoltaics in China and Germany," *Environmental Innovation and Societal Transitions* 17 (January 2015): 131.

17. Ibid., 134.
18. Ibid., 131.
19. Ibid., 132.
20. Ibid., 135.
21. Ibid., 136.
22. Ibid., 136, 138.
23. Ibid., 135.
24. Rainer Quitzow, "The Co-Evolution of Policy, Market and Industry in the Solar Energy Sector," FFU Report (Freie Universität Berlin, June 2013), 38.
25. Nagalakshmi Puttaswamy and Mohd Sahil Ali, "How Did China Become the Largest Solar PV Manufacturing Country?" Collegiate Science and Technology Entry Program (CSTEP) Working Paper 2, 2.
26. Ibid.
27. Quitzow, "Dynamics of a Policy Driven Market," 131; *China's Promotion of the Renewable Electric Power Industry: Hydro, Wind, Solar, Biomass*, Dewey & LeBoeuf LLP, National Foreign Trade Council, March 2010, 75; and Puttaswamy and Ali, "How Did China Become," 2.
28. Rainer Quitzow, cited in Donald Chung, Kelsey Horowitz, and Parthiv Kurup, "On the Path to Sunshot: Emerging Opportunities and Challenges in US Solar Manufacturing," National Renewable Energy Laboratory (NREL), U.S. Department of Energy, May 2016, 12.
29. "China 2008: The Global Financial Crisis," *China Digital Times*, December 8, 2008, http://chinadigitaltimes.net/2008/1 2/2008-financial-crisis-and-china/.
30. Jones and Bouamane, "'Power from Sunshine,'" 59.
31. Photovoltaic Power Systems Programme Annual Report 2014: "Implementing Agreement on Photovoltaic Power Systems," International Energy Agency, May 21, 2015, 62.
32. Jones and Bouamane, "'Power from Sunshine,'" 59.
33. Quitzow, "Dynamics of a Policy Driven Market," iii.
34. See, for example, Yuchao Zhu, "'Performance Legitimacy' and China's Political Adaptation Strategy," *Journal of Chinese Political Science* 16 (2011): 123–140.

35. Sufang Zhang, Philip Andrews-Speed, and Meiyun Ji, "The Erratic Path of the Low-Carbon Transition in China: Evolution of Solar PV Policy," *Energy Policy* 67, issue C (2014): 19.

36. Ibid.

37. Chen Gang, "China's Solar PV Manufacturing and Subsidies from the Perspective of State Capitalism," *Copenhagen Journal of Asian Studies* 33, no. 1 (2015): 96.

38. Ibid.

39. Ibid., 96–97.

40. Zhang, Andrews-Speed, and Ji, "The Erratic Path," 18.

41. Ibid., 23.

42. Gang, "China's Solar PV," 97.

43. Scott Neuman, "Chinese Solar Panel Maker Suntech Goes Bankrupt," NPR, March 20, 2013, https://www.npr.org/sections/thetwo-way/2013/03/20/174828432/chinese-solar-panel-maker-suntech-goes-bankrupt.

44. Ibid.

45. Gang, "China's Solar PV," 97.

46. Zhang, Andrews-Speed, and Ji, "The Erratic Path," 19.

47. Puttaswamy and Ali, "How Did China Become," 4.

48. John Fialka, "Why China Is Dominating the Solar Industry," *Scientific American*, December 19, 2016, https://www.scientificamerican.com/article/why-china-is-dominating-the-solar-industry/.

49. Eric Wesoff, "The Mercifully Short List of Fallen Solar Companies: 2015 Edition," Greentech Media, December 1, 2015, https://www.greentechmedia.com/articles/read/the-mercifully-short-list-of-fallen-solar-companies-2015-edition.

50. Jones and Bouamane, "'Power from Sunshine,'" 59.

51. Jones and Bouamane, "'Power from Sunshine,'" 69.

52. Gang, "China's Solar PV," 90.

53. Ibid, 90.

54. Jeffrey Ball, Dan Reicher, Xiaojing Sun, and Caitlin Pollock, *The New Solar System: China's Evolving Solar Industry and Its Implications for Competitive Solar Power in the United States and the World* (Stanford: Steyer-Taylor Center for Energy Policy and Finance, March 2017), 110.

55. Ibid.

CHAPTER 8

1. Ken Zweibel, quoted in John Fialka, "Why China Is Dominating the Solar Industry," *Scientific American*, December 19, 2016, https://www.scientificamerican.com/article/why-china-is-dominating-the-solar-industry/.

2. Luis Mundaca and Jessika Luth Richter, "Assessing 'Green Energy Economy' Stimulus Packages: Evidence from the U.S. Programs Targeting Renewable Energy," *Renewable and Sustainable Energy Reviews* 42 (2015), https://www.sciencedirect.com/science/article/pii/S1364032114008855.

3. Ibid.

4. Ibid.

5. Interestingly, this point about how different forms of government approach climate change echoes a disturbing theme of Naomi Oreskes and Erik M. Conway's book *The Collapse of Western Civilization: A View from the Future* (New York: Columbia University Press, 2014). Oreskes and Conway argue that today's persistent and fractured debates about the exercise of "big government" power to respond to climate change are, ironically, apt to increase the likelihood that, in time, crises brought on by climate change will precipitate strong, centralized government action necessary to protect life and property. The authors conclude, "And so the development that the neoliberals most dreaded—centralized government and loss of personal choice—was rendered essential by the very policies that they had put in place," (49).

6. "Renewable Energy and Jobs: Annual Review 2017," International Renewable Energy Agency, 13.

7. Ibid., 14.

8. SEIA Responds to Suniva's Section 201 Filing, Solar Energy Industries Association, April 27, 2017, https://www.seia.org/blog/seia-responds-sunivas-section-201-filing.

9. For instance, "The Chinese government, along with other countries, has used state subsidies and industrial strategies to advance its interests. America must accelerate its own R&D with a focus on developing the domestic supply chain for electric vehicles. A specific focus of Biden's historic R&D and procurement commitments will be on battery technology—for use in electric vehicles and on our

grid, as a complement to technologies like solar and wind—increasing durability, reducing waste, and lowering costs, all while advancing new chemistries and approaches. And Biden will ensure that these batteries are built in the United States by American workers in good, union jobs." Accessed November 15, 2020, https://joebiden.com/clean-energy/#

10. Gang, "China's Solar PV," 94, 100.

11. Ibid.

12. Ariana Eunjung Cha, "Solar Energy Firms Leave Waste Behind in China," *Washington Post*, March 9, 2008, http://www.washingtonpost.com/wp-dyn/content/article/2008/03/08/AR2008030802595.html.

13. Marcy Mason, "Solar Eclipse: How CBP Partnered with Industry to Save America's Solar Manufacturers," U.S. Customs and Border Protection, July 2015, https://www.cbp.gov/frontline/frontline-july-2015.

14. "Billion-Dollar Weather and Climate Disasters: Table of Events," National Centers for Environmental Information, accessed June 30, 2019, https://www.ncdc.noaa.gov/billions/events/US/1980–2019.

15. Ibid.

CHAPTER 9

1. Rachel Carson, *The Sea Around Us* (New York: Oxford University Press, 1951).

2. Scientist Eunice Foote conducted experiments in the 1850s that suggested that certain conditions, such as moist air and air with higher concentrations of carbon dioxide, led to warmer temperatures than those without. Her 1856 paper, "Circumstances Affecting the Heat of the Sun's Rays," stated, "An atmosphere of that gas (CO_2) would give to our earth a high temperature." See John Schwartz, "Overlooked No More: Eunice Foote, Climate Scientist Lost to History," *New York Times*, April 21, 2020. Svante Arrhenius researched the subject of CO_2 concentrations in the 1890s and wrote in 1908 that "the slight percentage of carbonic acid in the atmosphere may, by the advances of industry, be changed to a noticeable degree in the course of a few centuries." *Worlds in the Making: The Evolution of the Universe* (New York and London: Harper & Brothers Publish-

ers, 1908), 54.

3. Author's analysis of annual tomato production data published by the UN Food and Agriculture Organization (FAO), accessed May 15, 2019, http://www.fao.org/faostat/en/#data/QC.

4. "Fertilizer Out of Thin Air," BASF, accessed June 25, 2019, https://www.basf.com/us/en/media/science-around-us/fertilizer-out-of-thin-air.html

5. See Figure 1 in J. Poore et al., "Reducing Food's Environmental Impact Through Producers and Consumers," *Science* 360, no. 6392 (June 1, 2018): 987–92, accessed June 25, 2019, https://science.sciencemag.org/content/360/6392/987.full?ijkey=ffyeW1F0oS-l6k&keytype=ref&siteid=sci.

6. Bevan Griffiths-Sattenspiel and Wendy Wilson, "The Carbon Footprint of Water, A River Network Report," The Energy Foundation, 2009, https://www.csu.edu/cerc/researchreports/documents/CarbonFootprintofWater-RiverNetwork-2009.pdf.

7. Cynthia Stokes Brown, *Big History: From the Big Bang to the Present* (New York: The New Press, 2007), 42.

8. Max Roser and Esteban Ortiz-Ospina, "World Population Growth," Our World In Data, updated April 2017, https://ourworldindata.org/world-population-growth.

9. Evidence suggests that, in 1900–2200 BCE, coal was systematically used in certain areas of China, including Fushun, for heating and metallurgy work, but this use did not grow and diversify in the manner occurring later in England. John Dodson, Xiaoqiang Li, Nan Sun, Pia Atahan, Xinying Zhou, Hanbin Lie, Keliang Zhao, Songmei Hu, and Zemeng Yang, "Use of Coal in the Bronze Age in China," *The Holocene* 24, no. 5 (2014): 525–530.

10. Some would credit James Watt as a more significant figure than Newcomen in the history of the steam engine and Industrial Revolution. Reasonable cases can be made either way, but my own view is that Newcomen invented the idea while Watt modified Newcomen's, albeit in very important ways. Put differently, Watt didn't invent the steam engine, even though he may well have reinvented it. The exact degree to which Newcomen was in turn influenced by other inventions of his time appears to be lost to history. See Thomas Crump, *A Brief History of the Age of Steam* (Philadelphia:

Running Press, 2007), 51–56.

11. "Newcomen Atmospheric Engine," National Museum of Scotland, accessed May 6, 2018, https://www.nms.ac.uk/explore-our-collections/stories/science-and-technology/newcomen-engine?item_id=.

12. Crump, *A Brief History of the Age of Steam*, 53.

13. "Dynamo," Princeton University, accessed June 3, 2018, https://www.princeton.edu/ssp/joseph-henry-project/dynamo/Dynamo.pdf.

14. Maury Klein, *The Power Makers: Steam, Electricity, and the Men Who Invented Modern America* (New York: Bloomsbury Publishing, 2010), 252.

15. Ibid., 254.

16. Widespread use of mechanized farm equipment is well established in the United States and Russia. Regarding mechanized equipment in Kenya, while Sub-Saharan Africa has the lowest usage of agricultural mechanization in the world, motorized power is used in about 30 percent of Kenyan farms, particularly the mid- and large-size farm operations. See Noah W. Wawire et al., "The Status of Agricultural Mechanization in Kenya," July 2016, accessed August 2, 2018, http://www.kalro.org/sites/default/files/kafaci_report.pdf. Regarding China, that country's goal of 95 percent self-sufficiency in food grains by 2020 has buttressed a sustained push into agricultural mechanization, where 90 percent of wheat production is now mechanized, as is 60 percent of corn production and 40 percent of rice production. Current mechanization in China includes not just tractors, threshers, and the like but also drones distributing fertilizers and pesticides. Lilian Schaer, "China Powers Up," Country Guide, March 24, 2017, accessed August 2, 2018, https://www.country-guide.ca/2017/03/24/china-gets-serious-about-farm-machinery/50781/.

17. More than 87 percent of the world population had access to electricity in 2016. The vast majority of people without access to electricity are in rural areas of Sub-Saharan Africa, Myanmar, Cambodia, and North Korea. "Access to Electricity (percent of population)," The World Bank, accessed May 6, 2018, https://data.worldbank.org/indicator/EG.ELC.ACCS.ZS.

18. IEA World Energy Outlook 2002, Part II, 373–385 and 397,

accessed May 6, 2018, https://www.iea.org/media/weoweb-site/2008-1994/weo2002_part2.pdf.

19. Ivo Šlaus and Garry Jacobs, "Human Capital and Sustainability," *Sustainability* 1, no. 3 (2011): 97–154, http://www.mdpi.com/2071-1050/3/1/97/htm.

20. Max Roser, Hannah Ritchie, and Esteban Ortiz-Ospina, "World Population Growth," 2013, *OurWorldInData.org.* Accessed May 20, 2021, https://ourworldindata.org/world-population-growth. Historical population figures were taken from the same source at https://ourworldindata.org/world-population-growth#how-has-world-population-growth-changed-over-time.

21. It is more common to see the surging population and GDP growth from the late eighteenth century onward explained as a function of economic history—in particular, the reorientation of human interaction around the globe into capitalist and neo-capitalist behavior and the story of worldwide industrialization. I acknowledge that these views are soundly researched and supported and that I am unqualified academically to argue with historians who have authored such works. However, my own view is that, one hundred years from now, historians looking back through the lens of the dramatic events predicted to occur between now and then due to climate change will weigh the emissions aspect of the Industrial Revolution more heavily than the economic theory associated with it.

22. Christine Shearer, Nicole Ghio, Lauri Myllyvirta, Aiqun Yu, and Ted Nace, "Boom and Bust 2017: Tracking the Global Coal Plant Pipeline," Coalswarm/Sierra Club/Greenpeace, accessed May 6, 2018, http://endcoal.org/wp-content/uploads/2017/03/BoomBust2017-English-Final.pdf.

23. "Existing Nameplate and Net Summer Capacity by Energy Source, Producer Type and State (EIA-860)," U.S. Energy Information Administration, accessed May 6, 2018, https://www.eia.gov/electricity/data/state/.

24. Michael Forsythe, "China Cancels 103 Coal Plants, Mindful of Smog and Wasted Capacity," *New York Times,* January 18, 2017, https://www.nytimes.com/2017/01/18/world/asia/china-coal-power-plants-pollution.html?_r=0.

25. Lili Pike, "China Aims To Be Carbon Neutral by 2060. Its New 5-year Plan Won't Help," Vox, March 5, 2021, and Joe McDonald and Katy Daigle, "Report: Climate Outlook Improves as Fewer Coal Plants Built," Phys.Org, March 22, 2017, https://phys.org/news/2017-03-climate-outlook-coal-built.html.

26. Ottmar Edenhofer et al., "Reports of Coal's Terminal Decline May Be Exaggerated," *Environmental Research Letters* 13, no. 2 (February 7, 2018), http://iopscience.iop.org/article/10.1088/1748-9326/aaa3a2/meta.

27. Saleem Shaikh and Sughra Tunio, "Pakistan Ramps Up Coal Power with Chinese-Backed Plants," Reuters, May 2, 2017, https://www.reuters.com/article/us-pakistan-energy-coal/pakistan-ramps-up-coal-power-with-chinese-backed-plants-idUSKBN17Z019, and the U.S. Energy Information Administration website, https://www.eia.gov/electricity/state/.

28. Maria Avgerinou, Paolo Bertoldi, and Luca Castellazzi, "Trends in Data Centre Energy Consumption under the European Code of Conduct for Data Centre Energy Efficiency," accessed May 7, 2018, https://www.mdpi.com/1996-1073/10/10/1470/pdf.

29. "Energy and Climate Change," World Energy Outlook Special Report, International Energy Agency, accessed May 6, 2018, https://www.iea.org/publications/freepublications/publication/WEO2015SpecialReportonEnergyandClimateChange.pdf.

30. See, for instance, https://www.edf.org/climate/methane-studies.

31. "Understanding Global Warming Potentials," United States Environmental Protection Agency, accessed August 2, 2018, https://www.epa.gov/ghgemissions/understanding-global-warming-potentials.

CHAPTER 10

1. Christopher Hibbert, *The London Encyclopaedia*, 3rd ed. (London: Pan Macmillan, 2010), 248.

2. Ibid.

3. Giovanni De Feo et al., "The Historical Development of Sewers Worldwide," *Sustainability* 6 (2014): 3961, http://www.mdpi.com/2071-1050/6/6/3936/pdf.

4. Ibid., 3956.

5. Tim Evans, "Urban Drainage and the Water Environment: A Sustainable Future?" Foundation for Water Research, April 2013, 11, http://www.fwr.org/urbndnge.pdf.

6. Ibid.

7. "Cholera and London," Cholera and the Thames, accessed March 10, 2018, https://www.choleraandthethames.co.uk/cholera-in-london/cholera-in-westminster/.

8. Stephen Halliday, *The Great Stink of London: Sir Joseph Bazalgette and the Cleansing of the Victorian Metropolis* (Stroud, UK: The History Press, 1999).

9. "Cholera and Westminster," Cholera and the Thames, accessed July 28, 2018, https://www.choleraandthethames.co.uk/cholera-in-london/cholera-in-westminster/.

10. Halliday, *The Great Stink of London*.

11. "Seven Manmade Wonders: London Sewers," BBC homepage, England, September 24, 2014, http://www.bbc.co.uk/england/sevenwonders/london/sewers_mm/index.shtml.

12. Michael Kavanagh, "Super-Sewer Aims to Relieve Strain," *Financial Times*, October 30, 2011, https://www.ft.com/content/7e6d04f8-02ff-11e1-899a-00144feabdc0.

13. Karn Vohra, Alina Vodonos, Joel Schwartz, Eloise A. Marais, Melissa P. Sulprizio, Loretta J. Mickley, "Global Mortality from Outdoor Fine Particle Pollution Generated by Fossil Fuel Combustion: Results From GEOS-Chem," *Environmental Research* 195 (2021).

14. Rebecca McCarthy, "What a Pack of Cigarettes Costs in Every State," *The Awl*, July 11, 2017, https://www.theawl.com/2017/07/what-a-pack-of-cigarettes-costs-in-every-state-3/.

15. Lewis Strauss, "Remarks Prepared by Lewis L. Strauss, Chairman, United States Atomic Energy Commission, For Delivery at the Founders' Day Dinner, National Association of Science Writers, On Thursday, September 16, 1954," available at https://www.nrc.gov/docs/ML1613/ML16131A120.pdf

16. Douglas J. Arent et al., "Key Economic Sectors and Services," in *Climate Change 2014: Impacts, Adaptation, and Vulnerability—* Part A: Global and Sectoral Aspects (Cambridge and New York: Cambridge University Press, 2014), 665.

17. Clive Thompson, "Derailed," *New York Magazine*, February 2005, http://nymag.com/nymetro/news/features/11160/.

18. Niall R. McGlashan et al., "Negative Emissions Technologies," Imperial College, London, Grantham Institute for Climate Change, Briefing Paper no. 8, October 2012, https://www.imperial.ac.uk/media/imperial-college/grantham-institute/public/publications/briefing-papers/Negative-Emissions-Technologies-Grantham-BP-8.pdf; Eloy S. Sanz-Pérez et al., "Direct Capture of CO_2 from Ambient Air," *Chemical Reviews* 116, no. 19 (2016): 11840–76, http://pubs.acs.org/doi/10.1021/acs.chemrev.6b00173; and Elizabeth Kolbert, "Can Carbon-Dioxide Removal Save the World?" *The New Yorker*, November 20, 2017, https://www.newyorker.com/magazine/2017/11/20/can-carbon-dioxide-removal-save-the-world.

19. See, for instance, Jacob Bronsther, "The Climate Agreement in Lima Isn't Enough. Here's a Better Solution," *The New Republic*, December 15, 2014, https://newrepublic.com/article/120562/lima-climate-change-agreement-isnt-enough-time-air-capture.

20. France's remarkable Rance Tidal Power Station announced the arrival of tidal power when it was completed back in 1966 and has operated successfully and reliably ever since. At 240 megawatts, this plant is about half the size of an average coal plant. Unfortunately, the technology never caught on, although recently plants have been built in South Korea and China.

21. Joern Huenteler et al., "Why Is China's Wind Power Generation Not Living Up to Its Potential?" *Environmental Research Letters* 13, no. 4, March 19, 2018, accessed March 27, 2021, https://iopscience.iop.org/article/10.1088/1748-9326/aaadeb.

22. Ibid. The article concludes that underperformance of the wind projects' output was due to "delays in grid connection (14% of the gap) and curtailment due to constraints in grid management (10% of the gap)" and "suboptimal turbine model selection (31% of the gap), wind farm siting (23% of the gap), and turbine hub heights (6% of the gap)."

23. Devin Henry, "Dem Senator: There Are 'Deniers' on Both Sides of Climate Change," *The Hill*, August 8, 2016, http://thehill.com/policy/energy-environment/275682-dem-senator-there-are-deniers-on-both-sides-of-climate-change.

24. "Reactors Are Closing," Beyond Nuclear, http://www.beyondnu-clear.org/reactors-are-closing/.

25. Andrew Follett, "First US Nuclear Reactor in 20 Years Is Fully Operational," *The Daily Caller*, October 21, 2016, http://dailycaller.com/2016/10/21/first-us-nuclear-reactor-in-2 0-years-is-fully-operational/.

26. Michael Hiltzik, "America's First '21st Century Nuclear Plant' Already Has Been Shut Down for Repairs," *Los Angeles Times*, May 8, 2017, http://www.latimes.com/business/hiltzik/la-fi-hiltzik-nuclear-shutdown-20170508-story.html.

27. Peter Maloney, "Westinghouse Bankruptcy Could Grind US Nuclear Sector to a Halt," Utility Dive, April 12, 2017, https://www.utilitydive.com/news/westinghouse-bankruptc y-could-grind-us-nuclear-sector-to-a-halt/440153/, and Dave Williams, "Plant Vogtle Opponents Seeking Faster Review of Cost to Customers," *The Augusta Chronicle*, August 16, 2020.

28. Ibid.

29. Peter Maloney, "Toshiba Said to Be Preparing for Bankruptcy as Southern Faces Vogtle Deadline," Utility Dive, May 12, 2017, https://www.utilitydive.com/news/toshiba-said-to-be-preparing-fo r-bankruptcy-as-southern-faces-vogtle-deadli/442588/.

30. "Analytical Perspectives: Budget of the U.S. Government," Fiscal Year 2018, Office of Management and Budget, 2017, 216, https://www.whitehouse.gov/sites/whitehouse.gov/files/omb/budget/fy2018/spec.pdf.

31. "The Budget and Economic Outlook: 2017 to 2027," Congressional Budget Office, January 2017, https://www.cbo.gov/sites/default/files/115th-congress-2017-2018/reports/52370-outlookonecolumn.pdf.

32. Peter Howard and Derek Sylvan, "Expert Consensus on the Economics of Climate Change," New York University School of Law Institute for Policy Integrity, December 2015, http://policyintegrity.org/files/publications/ExpertConsensusReport.pdf.

33. OECD, *The Economic Consequences of Climate Change* (Paris: OECD Publishing, 2015), 3, http://dx.doi.org/10.1787/9789264235410-en.

34. Å Johansson et al., *Looking to 2060: Long-Term Global Growth*

Prospects, OECD Economic Policy Paper Series (Paris: OECD Publishing, November 2012, https://www.oecd.org/eco/outlook/2060%20policy%20paper%20FINAL.pdf.

35. OECD, *Environmental Outlook to 2050* (Paris: OECD Publishing, 2012), 49. See also numerous other pre-2015 OECD reports on climate change, such as *Climate Change and Agriculture: Impacts, Adaptation and Mitigation* (2010), *Cities and Climate Change* (2010), and *The Economics of Climate Change Mitigation: Policies and Options for Global Action Beyond 2012* (2009).

36. United Nations, Department of Economic and Social Affairs, Population Division (2019). World Population Prospects 2019: Highlights, https://population.un.org/wpp/Publications/Files/WPP2019_Highlights.pdf.

37. Ibid., 14.

38. Ibid.

39. I. Niang et al., "Africa," in *Climate Change 2014: Impacts, Adaptation, and Vulnerability—Part B: Regional Aspects*. Contribution of Working Group II to the Fifth Assessment Report of the Intergovernmental Panel on Climate Change (Cambridge and New York: Cambridge University Press, 2014), 1199–1265.

40. Y. Hijioka, et al., "Asia," in *Climate Change 2014: Impacts, Adaptation, and Vulnerability. Part B: Regional Aspects*. Contribution of Working Group II to the Fifth Assessment Report of the Intergovernmental Panel on Climate Change (Cambridge and New York: Cambridge University Press, 2014), 1327–70.

41. "The Fragile States Index (FSI) produced by The Fund for Peace (FFP) is a critical tool in highlighting not only the normal pressures that all states experience, but also in identifying when those pressures are outweighing a states' capacity to manage those pressures. By highlighting pertinent vulnerabilities which contribute to the risk of state fragility, the Index—and the social science framework and data analysis tools upon which it is built—makes political risk assessment and early warning of conflict accessible to policy-makers and the public at large." *Fragile States Index Annual Report 2020*, 39, https://fragilestatesindex.org/wp-content/uploads/2020/05/fsi2020-report.pdf.

42. Andreas Malm, *Fossil Capital: The Rise of Steam Power and the*

Roots of Global Warming (Brooklyn: Verso Books, 2016), 381–95.

43. Carbon emissions data taken from the Global Carbon Atlas and available at http://www.globalcarbonatlas.org/en/CO2-emissions, accessed July 29, 2018.

44. Margins of 22,748 votes in Wisconsin, 44,307 votes in Pennsylvania, and 10,704 votes in Michigan pushed these states' Electoral College votes into the Trump column, and their collective forty-six Electoral College votes made the difference in the outcome. In contrast, Clinton won California by about four million votes and New York by 1.5 million.

45. Dante Chinni, "Did Biden Win by a Little or a Lot? The Answer Is . . . Yes," NBC News, December 20, 2020, accessed March 27, 2021, https://www.nbcnews.com/politics/meet-the-press/did-biden-win-little-or-lot-answer-yes-n1251845

46. The Climate Stewardship Act enjoyed the support of six Republican senators. The bill was ultimately defeated in a vote of 55–43, with ten Democratic senators voting against.

47. James Madison, Session of Thursday, July 19, 1787, in The Debates in the Federal Convention of 1787 Which framed the Constitution of the United States, 282, 285–86, quoted in Juan Perea, "Echoes of Slavery II: How Slavery's Legacy Distorts Democracy," University of California at Davis Journal, 1088.

48. See, for instance, Allen Guelzo and James Hulme, "In Defense of the Electoral College," *Washington Post*, November 15, 2016. "Above all, the electoral college had nothing to do with slavery. Some historians have branded the electoral college this way because each state's electoral votes are based on that 'whole Number of Senators and Representatives' from each State, and in 1787 the number of those representatives was calculated on the basis of the infamous 3/5ths clause. But the electoral college merely reflected the numbers, not any bias about slavery (and in any case, the 3/5ths clause was not quite as proslavery a compromise as it seems, since Southern slaveholders wanted their slaves counted as 5/5ths for determining representation in Congress, and had to settle for a whittled-down fraction). As much as the abolitionists before the Civil War liked to talk about the 'proslavery Constitution,' this was more of a rhetorical posture than a serious historical argument. And the simple

fact remains, from the record of the Constitutional Convention's proceedings (James Madison's famous Notes), that the discussions of the electoral college and the method of electing a president never occur in the context of any of the convention's two climactic debates over slavery."

CHAPTER 11

1. As reported on http://www.hurricanecity.com/city/galveston.htm, accessed July 5, 2019.

2. Neena Satija, Kiah Collier, Al Shaw, Jeff Larson, "Hell and High Waters," ProPublica, accessed June 26, 2019, https://projects.pro-publica.org/houston/#.

3. Jeannie Ralston, "Ike Dike to the Rescue," Texas A&M Founda-tion, accessed June 26, 2019, https://www.txamfoundation.com/News/Ike-Dike-to-the-Rescue.aspx, and Diana Budds, "Ike Dike: The $15 Billion Storm Surge Barrier Houston Can't Agree On," Fast Company, accessed June 26, 2019, https://www.fastcom-pany.com/90138474/ike-dike-the-15-billion-storm-surge-barrie r-houston-cant-agree-on.

4. Ibid.

5. Mary Beth Griggs, "Hurricanes Are Getting More Intense—But Should We Add a Category 6?" *Popular Science*, accessed June 26, 2019, https://www.popsci.com/category-6-hurricane.

6. Kiah Collier, "Army Corps Set to Propose Hurricane Protection Plan for Houston," *Texas Tribune*, accessed June 26, 2019, https://www.texastribune.org/2018/10/03/army-corps-set-propose-hurrica ne-protection-plan-houston/.

7. Julian Gill, "Galveston Seawall, Other Cities Will Be Underwater by 2100, According to Sea Level Simulator," *Houston Chronicle*, June 19, 2019.

8. See, for instance, "Final Report: Integrated Hurricane Sandy General Reevaluation Report and Environmental Impact State-ment," US Army Corps of Engineers, May 2019, accessed July 3, 2019, https://www.nan.usace.army.mil/Portals/37/docs/civilworks/projects/ny/coast/Rockaway/Rockaway%20Final%20Report/Rock%20Jam%20Bay%20Final%20Report%20HSGRR%205-9-19.pdf?ver=2019-05-29-124532-717; and "Protecting Boston: The

Boston Harbor Barrier Study," EBC Climate Change Program, 2018, accessed July 3, 2019, https://ebcne.org/wp-content/upload s/2018/06/06-22-18-MASTER-Climate-Change-Program-The-B oston-Harbor-Barrier-Study.pdf.

9. See statement regarding "Lake Pontchartrain and Vicinity General Re-Evaluation Report," US Army Corps of Engineers, accessed July 3, 2019, https://www.mvn.usace.army.mil/About/Projects/ BBA-2018/studies/LPV-GRR/.

10. Ibid. According to the US Army Corps of Engineers, "Engineering analysis indicates the HSDRRS [Hurricane & Storm Damage Risk Reduction System] will no longer provide 1% level of risk reduction as early as 2023. Absent future levee lifts to offset consolidation, settlement, subsidence, and sea level rise, risk to life and property in the Greater New Orleans area will progressively increase."

11. Knowledgeable historians contend that it may not have been Winston Churchill who said this. Instead, the quote may be attributable to Israeli politician Abba Eban, in 1967. See http://quoteinvestigator.com/2012/11/11/exhaust-alternatives/ and http://www.npr.org/sections/itsallpolitics/2013/10/28/241295755/a-churchill-quote-that-u-s-politicians-will-never-surrender.

12. "Timeline of Polling History: Events That Shaped the United States, and the World," Gallup, accessed July 7, 28, 2018, http://news.gallup.com/poll/9967/timeline-polling-history-event s-shaped-united-states-world.aspx, and Adam J. Berinsky, *In Time of War: Understanding American Public Opinion from World War II to Iraq* (Chicago: University of Chicago Press, 2009).

13. Lydia Saad and Jeffrey M. Jones, "U.S. Concern About Global Warming at Eight-Year High," Gallup, March 16, 2016, http://news.gallup.com/poll/190010/concern-global-warming-eight-year-high.aspx?g_source=CATEGORY_CLIMATE_CHANGE&g_medium=topic&g_campaign=tiles; "Fact Sheet—Polling the American Public on Climate Change (2015)," Environmental and Energy Study Institute, April 10, 2015, http://www.eesi.org/papers/view/ fact-sheet-polling-the-american-public-on-climate-change-2015; and "Fact Sheet—Polling the American Public on Climate Change (2014)," Environmental and Energy Study Institute, October 14, 2014, http://www.eesi.org/papers/view/fact-sheet-polling-

the-american-public-on-climate-change-2014.

14. "Presidential Historians Survey 2017," C-SPAN, https://
www.c-span.org/presidentsurvey2017/?page=overall.

15. "The 100 Most Influential Americans of All Time," *The Atlantic*,
December 2006.

EPILOGUE

1. Michael J. Coren, "Electric Airplanes Are Getting Tantalizingly
Close to a Commercial Breakthrough," Quartz, December 13,
2021, https://qz.com/1943592/electric-airplanes-are-gettin
g-close-to-a-commercial-breakthrough/, and Kevin Bullis, "Once
a Joke, Battery-Powered Airplanes Are Nearing Reality," *MIT
Technology Review*, July 8, 2013, https://www.technologyreview.
com/s/516576/once-a-joke-battery-powered-airplanes-are-nearing-
reality/. "The batteries also make it possible to recover energy
during descent much the way hybrid cars capture energy during
braking (propellers spin a generator). And, as batteries improve,
they will provide more and more of the energy on board."

2. "Water captured in reservoirs or pumped from faraway deltas is
getting more expensive—and such alternatives come with their own
environmental costs. As sources dry up and competition for water
mounts from businesses, farmers, and cities, we will inevitably turn
to seawater and other salty sources. It might not be a great solution, but the bottom line is that we are left with fewer and fewer
choices in a water-starved world." David Talbot, "Desalination
Out of Desperation," *MIT Technology Review*, December 16, 2014,
https://www.technologyreview.com/s/533446/desalination-ou
t-of-desperation/.

3. "Many of Southern California's beaches . . . could retain their width
if they were allowed to expand inland, which is what would happen
naturally if highways, buildings, sewer pipes and other artifacts of
development weren't in the way. Beach house owners and local governments will resist abandoning or relocating their structures, but
if no action is taken, the beaches will shrivel and the property will
get flooded out anyway." Jacques Leslie, "California Faces a Cascade
of Catastrophes as Sea Level Rises," *Los Angeles Times*, January
24, 2018, http://www.latimes.com/opinion/op-ed/la-oe-leslie-se

a-level-rise-california-20180124-story.html.

4. "Across Mexico, farmers still wait for rain that doesn't come. Severe droughts, punctuated by intense storms and flooding, are huge environmental challenges for Mexico in the coming century. By 2080, agricultural declines are expected to drive 1.4 million to 6.7 million adult Mexicans out of the country. Most of those people will come to the United States." Amy McDermott, "Mexico's climate migrants are already coming to the United States," Grist, December 29, 2016, http://grist.org/article/mexicos-climate-migrants-are-already-coming-to-the-united-states/.

5. See Chisaki Watanabe, "Researchers Think They're Getting Closer to Making Spray-On Solar Cells a Reality," Bloomberg, March 20, 2017, https://www.bloomberg.com/news/articles/2017-03-21/the-wonder-material-that-may-make-spray-on-solar-cells-a-reality; "Anita Ho-Baillie's Team Has Set a New World Efficiency Record for Trend-Setting Solar Cells," UNSW School of Photovoltaic and Renewable Energy Engineering, January 24, 2017, https://www.engineering.unsw.edu.au/energy-engineering/news/spray-on-solar-cells; and James Randerson, "Spray On and Printable: What's Next for the Solar Panel Market," *Guardian*, May 4, 2017, https://www.theguardian.com/sustainable-business/2017/may/04/solar-renewables-energy-thin-film-technology-perovskite-cells.

6. The federal government owns 47 percent of lands in the Western United States. Quoctrung Bui and Margot Sanger-Katz, "Why the Government Owns So Much Land in the West," *New York Times*, January 5, 2016.

7. Jim Robbins, "On the Water-Starved Colorado River, Drought Is the New Normal," YaleEnvironment360, January 22, 2019, https://e360.yale.edu/features/on-the-water-starved-colorado-river-drought-is-the-new-normal. "If climate change is locked in, [University of Michigan researcher Jonathan Overpeck] said, what is going on now is not a new normal, but a stop along the way to a yet-drier new normal somewhere in the distant future. . . . That's why the alarm is palpable among water managers in the Southwest. They see the writing on the wall, and there are few skeptics about climate change among them. The plight of Cape Town, South Africa,

which came to the brink of a water system crash last year, is on many people's minds along the Colorado River. This era of drying is especially serious because so much--some 40 million people and an economy that includes the world's fifth largest, in California--is riding on the flow of the Colorado. The specter of a region facing an existential crisis because of a warming climate becomes more real every day. 'If you can see it, you should plan for it,' [director of the Phoenix Water Services Department Kathryn] Sorensen said. 'And I can see it.'"

8. See Tony Davis, "Lake Powell Could Dry Up in as Little as Six Years, Study Says," *Arizona Daily Star*, September 3, 2016, http://tucson.com/news/local/lake-powell-could-dry-up-in-as-little-as-six/article_0e0b61d1-10d7-51aa-a29a-bcb1d9d6eb0f.html; Brandon Loomis, "Arizona Will Suffer More Than Most of U.S. as Climate Changes," *Arizona Republic*, June 29, 2017, https://www.azcentral.com/story/news/local/arizona-environment/2017/06/29/study-predicts-rising-heat-deaths-costs-arizona-climate-change/439573001/; and Mike Pearl, "Phoenix Will Be Almost Unlivable by 2050 Thanks to Climate Change," Vice, September 18, 2017, https://www.vice.com/en_us/article/vb7mqa/phoenix-will-be-almost-unlivable-by-2050-thanks-to-climate-change.

9. Former coal plant cooling towers have already been deployed for this purpose in South Africa, legally (see https://www.gauteng.net/attractions/orlando_towers) and in Russia by trespassers (see http://www.dailymail.co.uk/news/article-2089509/The-amazing-video-bungee-jumpers-leaping-Russian-cooling-tower-kept-safe-garden-hose.html).

10. Damon Lavrinc, "Elon Musk Thinks He Can Get You from NY to LA in 45 Minutes," *Wired*, July 15, 2013, https://www.wired.com/2013/07/elon-musk-hyperloop/.

11. "While most models project continued, long-term declines in lake levels, shorter-term variations will remain large, and periods of high lake levels are probable." See "Climate Change in the Great Lakes Region," Great Lakes Integrated Sciences Assessments, last updated June 18, 2014, http://glisa.umich.edu/media/files/GLISA_climate_change_summary.pdf.

12. See Kevin O'Sullivan, "Google, Amazon, and Apple 'Set to Enter'

Energy Supply Business," *Irish Times*, May 16, 2017, https://
www.irishtimes.com/news/environment/google-amazon-and-appl
e-set-to-enter-energy-supply-business-1.3085594; Lauren Hepler,
"Why Apple's New Energy Business Should Scare Utilities," Green
Biz, June 14, 2016, https://www.greenbiz.com/article/why-apple
s-new-energy-business-should-scare-utilities; and Davide Savenije,
"Is Google Becoming an Energy Company?" Utility Dive, January
23, 2014, https://www.utilitydive.com/news/is-google-becoming-a
n-energy-company/216848/.

13. Jeff McMahon, "How Gay Marriage Suggests a Strategy for Climate Change," *Forbes*, February 15, 2017, https://www.forbes.com/
sites/cdw/2017/09/21/how-iot-is-making-workplaces-more-effi-
cient/#57a0ccf16ef0.

14. "The Utilities sector is most exposed to carbon costs, with almost
half of EBITDA [earnings before interest, taxes, depreciation, and
amortization] at risk across the sector, depending on the ability
of companies to pass costs through to customers or switch to
low-carbon power generation. Within the sector, carbon costs as
a percentage of EBITDA at a company level ranges from 2% for
PG&E [Pacific Gas and Electric] Corporation to 117% for American Electric Power. . . . For three companies, carbon costs could
wipe out all earnings: Allegheny Energy, American Electric Power
and Ameren Corporation." "Carbon Risks and Opportunities in
the S&P 500," Trucost, Commissioned by Investor Responsibility Research Center Institute for Corporate Responsibility, June
2009, 27, https://irrcinstitute.org/wp-content/uploads/2015/09/
irrc_trucost_09061.pdf.

15. Sam Shead, "DeepMind Is Funding Climate Change Research at
Cambridge As It Looks to Use AI to Slow Down Global Warming," *Business Insider*, June 21, 2017, http://www.businessinsider.
com/deepmind-is-funding-climate-change-research-at-cambr
idge-university-2017-6.

16. "Ocean waves and tidal movements hold vast amounts of energy.
The US National Renewable Energy Laboratory estimates that
more than one third of all electricity used in the United States could
be drawn from the seas." Sophia Schweitzer, "Will Tidal and Wave
Energy Ever Live Up to Their Potential?" Yale Environment 360,

October 15, 2015, https://e360.yale.edu/features/will_tidal_and_wave_energy_ever_live_up_to_their_potential.

17. According to Richard Seager, co-author of a study linking the Syrian conflict to climate change, "We used to be rather skeptical about the domino theory as it was applied to Communist states in the 1960s and 1970s, but here we're definitely seeing a domino effect that began with invasions of Iraq and Afghanistan and the way that that's been rippling from one country to another. Maybe the dominoes just became easier to topple over once you add climate change into the mix." As quoted in Lucy Westcott, "Climate Change Helped Create Conditions for War in Syria," *Newsweek*, March 4, 2015, http://www.newsweek.com/climate-change-helped-create-conditions-war-syria-study-suggests-311199. For the full study, see Colin P. Kelley et al., "Climate Change in the Fertile Crescent and Implications of the Recent Syrian Drought," *PNAS* 112, no. 11 (March 17, 2015): 3241–46.

18. Rachel Thompson, "This Berlin Supermarket Just Installed a Farm Inside Its Store," Mashable, April 7, 2016, https://mashable.com/2016/04/07/supermarket-farm-berlin/#dTX6nlIzp5qF.

19. "California agriculture is uniquely vulnerable to climate change. Rising temperatures, constrained water resources, and increased pest and disease pressure are among the climate change impacts that threaten to fundamentally challenge California agriculture in the coming years and decades." See "Climate Change Impacts on Agriculture" fact sheet, California Climate and Agriculture Network, September 2011, http://calclimateag.org/wp-content/uploads/2011/09/Impacts-fact-sheet.pdf. See also http://www.climatechange.ca.gov/adaptation/documents/Statewide_Adaptation_Strategy_-_Chapter_8_-_Agriculture.pdf.

20. "China's water scarcity and its widening north-south water gap have increased pressure to execute controversial water diversion plans. These plans will threaten India, especially since the Brahmaputra River flows through a disputed area. These factors, plus changing domestic conditions in China, may increase the likelihood of war." Jin H. Pak, "China, India, and War over Water," 2016 Strategic Studies Institute, http://ssi.armywarcollege.edu/pubs/parameters/issues/Summer_2016/8_Pak.pdf.

21. "If we start opening the floodgates on some of these glaciers, even if we

stop our emissions, even if we go back to a better climate, the damage is going to be done. There's no red button to stop this." Elizabeth Kolbert, "Greenland Is Melting," *New Yorker*, October 24, 2016, https://www.newyorker.com/magazine/2016/10/24/greenland-is-melting.

22. J. C. Acosta Navarro et al., "Century Corrigendum: Amplification of Arctic Warming by Past Air Pollution Reductions in Europe," *Geoscience* 9, (2016): 277–81 [published online March 14, 2016; corrected after print May 5, 2016], https://www.nature.com/articles/ngeo2673, and Chelsea Harvey, "How Cleaner Air Could Actually Make Global Warming Worse," *Washington Post*, March 14, 2016, https://www.washingtonpost.com/news/energy-environment/wp/2016/03/14/how-cleaner-air-could-actually-make-global-warming-worse/?utm_term=.eaf9e8501043.

AFTERWORD

1. See for example National Centers for Environmental Information, "Assessing the Global Climate in August 2023," August 2023, accessed September 18, 2023, https://www.ncei.noaa.gov/news/global-climate-202308; and "Summer 2023: The Hottest on Record," Copernicus Climate Change Service, accessed September 18, 2023, https://climate.copernicus.eu/summer-2023-hottest-record.

2. Maya Yang, "Phoenix Sets Record in Hellishly Hot Summer—But Relief Is in Sight," *The Guardian*, September 2023, accessed September 25, www.theguardian.com/world/2023/sep/10/phoenix-arizona-new-heat-record-hot-summer.

3. Ian Livingston, "Phoenix Will Be First Major City to Average 100+ Degrees All Summer," *Washington Post*, July 21, 2023, accessed August 8, 2023, https://www.washingtonpost.com/weather/2023/07/21/phoenix-average-monthly-temperature-100-degrees/, and Jonathan Erdman, "Phoenix Had a Record Hottest Month for Any US City. Dozens of Cities Also Had Their Hottest Month," August 2, 2023, accessed September 13, 2023, Weather Underground, https://www.wunderground.com/article/safety/heat/news/2023-07-31-hottest-month-record-july-2023-phoenix-us-cities.

4. The suite of most recent UN IPCC reports, titled the "Sixth

Assessment Report," can be found at https://www.ipcc.ch/
assessment-report/ar6/.

5. See article at https://www.scientificamerican.com/article/this-ho
t-summer-is-one-of-the-coolest-of-the-rest-of-our-lives/.

6. National Ocean and Atmospheric Administration (NOAA), "The
Ongoing Marine Heat Waves in U.S. Waters, Explained," July 14,
2023, accessed September 15, 2023, https://www.noaa.gov/news/
ongoing-marine-heat-waves-in-us-waters-explained#.

7. NOAA, "Global Ocean Roiled by Marine Heatwaves, With More
on the Way," June 28, 2023, accessed September 4, 2023, https://
research.noaa.gov/2023/06/28/global-ocean-roiled-by-marine-he
atwaves-with-more-on-the-way/, and https://psl.noaa.gov/
marine-heatwaves/#report.

8. Dan Stillman, "Florida Ocean Temperatures at 'Downright Shocking' Levels," *Washington Post*, July 10, 2023, accessed August 4,
2023, https://www.washingtonpost.com/weather/2023/07/10/
florida-ocean-temperature-heat-records/, https://www.npr.
org/2023/07/17/1188042802/climate-scientists-are-alarme
d-by-record-water-temperatures-off-of-floridas-coas, and https://
www.bbc.com/news/science-environment-65948544.

9. Alan Urban, "2023 Is the Year Climate Change Went Exponential," *Medium*, July 2023, accessed September 14, 2023,
https://medium.com/@CollapseSurvival/2023-is-the-year-cli-
mate-change-went-exponential-88891ef68e88.

10. "It's Not Just Climate Change: Three Other Factors Driving
This Summer's Extreme Heat," *Yale Environment 360*, July 28,
2023, accessed September 26, 2023, https://e360.yale.edu/digest/
summer-2023-extreme-heat-causes.

11. "El Nino to Return This Year, Boosting Warming Trend," *Yale
Environment 360*, May 3, 2023, accessed September 26, 2023,
https://e360.yale.edu/digest/el-nino-2023-climate-change.

12. Li Yuan, "Strong El Nino Expected to Drive Record-Breaking
Global Surface Temperatures and Trigger Climate Crises in 2023–
24," *Phys.org*, September 28, 2023, accessed September 30,
2023, https://phys.org/news/2023-09-strong-el-nio-record-
breaking-global.html; and Kevin Li et al., "Record-Breaking
Global Temperature and Crises with Strong El Nino in 2023–

24," *The Innovation Geoscience*, 2023, accessed September 30, 2023, https://www.the-innovation.org/article/doi/10.59717/j.xinn-geo.2023.100030.

13. See IEA, "Renewable Energy Market Update - June 2023," Paris, https://www.iea.org/reports/renewable-energy-market-update-june-2023, License: CC BY 4.0.

14. "China's New Coal Plant Approvals Surge in 2022, Highest Since 2015," Reuters, February 26, 2023, accessed September 11, 2023, https://www.reuters.com/world/asia-pacific/chinas-new-coal-plant-approvals-surge-2022-highest-since-2015-research-2023-02-27/.

15. Keith Bradsher, "Why Heat Waves Are Deepening China's Addiction to Coal," July 20, 2023, accessed August 18, 2023, coahttps://www.nytimes.com/2023/07/20/business/china-coal-climate-change.html.

16. Sarita Singhe, "India, China Propose 'Multiple Pathways' on Cutting Use of Fossil Fuels," May 2, 2023, accessed June 19, 2023, https://www.reuters.com/world/india-china-propose-multiple-pathways-cutting-use-fossil-fuels-sources-2023-05-02/.

17. Matthew Chye and Carman Chew, "India's Power Output Grows at Fastest Pace in 33 Years, Fuelled by Coal," Reuters, April 5, 2023, accessed June 5, 2023, https://www.reuters.com/business/energy/indias-power-output-grows-fastest-pace-33-years-fuelled-by-coal-2023-04-05/.

18. Gavin Maquire, "India's Pledge to Stop New Coal Power Plants to Hit Key States," Reuters, May 9, 2023, accessed June 5, 2023, https://www.reuters.com/markets/commodities/indias-pledge-stop-new-coal-power-plants-hit-key-states-2023-05-09/.

19. Vera Eckert and Tom Sims, "Energy Crisis Fuels Coal Comeback in Germany," Reuters, December 2015, 2022, accessed May 30, 2023, https://www.reuters.com/markets/commodities/energy-crisis-fuels-coal-comeback-germany-2022-12-16/.

20. Benjamin Storrow, "The U.S. Can Hit Its 2030 Climate Goals, but a Lot Has to Go Right," *Scientific American*, March 30, 2023, accessed May 20, 2023, https://www.scientificamerican.com/article/the-u-s-can-hit-its-2030-climate-goals-but-a-lot-has-to-go-right/.

21. Secretary General, United Nations, "Secretary-General Warns of Climate Emergency, Calling Intergovernmental Panel's Report 'a

File of Shame,' While Saying Leaders 'Are Lying,' Fueling Flames,'" press release, accessed June 11, 2023, https://press.un.org/en/2022/sgsm21228.doc.htm.

22. IEA, World Energy Outlook 2022, Paris, https://www.iea.org/reports/world-energy-outlook-2022, License: CC BY 4.0 (report); CC BY NC SA 4.0 (Annex A).

23. Ibid.

24. Matteo Wong, "The Internet's Next Great Power Suck," *The Atlantic*, August 23, 2023, accessed September 11, 2023, https://www.theatlantic.com/technology/archive/2023/08/ai-carbon-emissions-data-centers/675094/.

25. See P. R. Shukla, J. Skea, R. Slade, A. Al Khourdajie, R. van Diemen, D. McCollum, M. Pathak, S. Some, P. Vyas, R. Fradera, M. Belkacemi, A. Hasija, G. Lisboa, S. Luz, J. Malley, (eds.), *Climate Change 2022: Mitigation of Climate Change. Contribution of Working Group III to the Sixth Assessment Report of the Intergovernmental Panel on Climate Change*, IPCC (New York: Cambridge University Press, 2022), 62, doi: 10.1017/9781009157926.

26. P. M. Forster, et al., "Indicators of Global Climate Change 2022: Annual Update of Large-Scale Indicators of the State of the Climate System and Human Influence," *Earth System Science Data* 15 (2023): 2295–2327, https://doi.org/10.5194/essd-15-2295-2023.

27. Caroline Hickman, Elizabeth Marks, Panu Pihkala, Susan Clayton, R. Eric Lewandowski, Elouise E. Mayall, Britt Wray, Catriona Mellor, Lise van Susteren, "Climate Anxiety in Children and Young People and Their Beliefs About Government Responses to Climate Change: A Global Survey," *The Lancet Planetary Health* 5, no. 12 (2021): e863–e873, ISSN 2542-5196, https://doi.org/10.1016/S2542-5196(21)00278-3.

28. Abby Kiesa, "Near Record-High Numbers of Young People Voted During the Midterms," *New Hampshire Bulletin*, December 29, 2022, accessed September 20, 2023, https://newhampshire-bulletin.com/2022/12/29/near-record-high-numbers-of-young-people-voted-during-the-midterms/.

29. Brett Christophers, "Why Are We Allowing the Private Sector to Take Over Our Public Works," *New York Times*, May 8, 2023, accessed June 18, 2023, https://www.nytimes.com/2023/05/08/

opinion/inflation-reduction-act-global-asset-managers.html; and
Brian Alexander, "Privatizing Is Changing America's Relationship
with Its Physical Stuff," *The Atlantic*, July 12, 2017, accessed May
14, 2023, https://www.theatlantic.com/business/archive/2017/07/
infrastructure-private-public-partnerships/533256/.

30. See Avery Ellfelt, "BlackRock Tallied Its Climate Impact. Here's
What It Found," Politico, February 1, 2022, accessed June 4, 2023,
https://www.politico.com/news/2022/02/01/wall-street-giant-
climate-impact-blackrock-00003447; and Union of Concerned
Scientists, "Each Country's Share of CO_2 Emissions, published July
16, 2008, updated January 14, 2022," accessed June 4, 2023, https://
www.ucsusa.org/resources/each-countrys-share-co2-emissions.

31. Tim Malton, "The Hidden Cause of the Titanic Disaster: Thermal Inversion and the Titanic," HistoryHit, June 29, 2020,
accessed April 9, 2023, https://www.historyhit.com/titanic-disaste
r-thermal-inversion/.

32. Ibid.

33. Ibid. See also David Bressan, "The Climate Science Behind the
Sinking of the Titanic," April 12, 2017, *Forbes*, accessed April 9,
2023, https://www.forbes.com/sites/davidbressan/2017/04/12/
the-climate-science-behind-the-sinking-of-the-titanic/; and Grant
R. Bigg, "Iceberg Risk in the Titanic Year of 1912: Was It Exceptional?" Royal Meteorological Society, *Weather* 69, no. 4, April
10, 2014, 100–104, which reaches a slightly different conclusion:
"In 1912, the peak number of icebergs for the year was recorded in
April, whereas normally this occurs in May, and there were nearly
two and a half times as many icebergs as in an average year. . . . The
year 1912 was indeed unusual, with 1,038 icebergs observed to cross
48N. However, this number does not even reach the 90th percentile of the annual number distribution . . . 1912 was a significant
ice-year, but not extreme."

34. N. Abram, H. McGregor, J. Tierney, et al., "Early Onset
of Industrial-Era Warming Across the Oceans and Conti-
nents," *Nature* 536 (2016): 411–418, https://doi.org/10.1038/
nature19082.

35. Ibid.

SUGGESTED READING

Brannen, Peter. *The Ends of the World: Volcanic Apocalypses, Lethal Oceans, and Our Quest to Understand Earth's Past Mass Extinctions*. New York: HarperCollins, 2017.

> While my own book focuses on the role and plight of human beings in the story of climate change, Brannen's book covers the major extinction-level events that have been visited upon the earth in all of history and in doing so provides a much wider vantage on how runaway climate change might affect all species and earth systems. Brannen's book is also effective at putting whatever happens with climate change into perspective in the much larger story of our planet.

Brown, Cynthia Stokes. *Big History: From the Big Bang to the Present*. New York: The New Press, 2007.

> Among my most recommended books on climate change, this book comes to the topic only briefly but does more to frame and reveal the dynamics behind it than anything else I have read. A joy to read, it served as the history book I wish I had been taught in high school due to its elegant, grounded telling

of the whole of history in an easily accessible and enjoyable session, as though you were sitting and talking with the whole *Encyclopaedia Britannica*.

Carson, Rachel. *The Sea Around Us*. New York: Oxford University Press, 1951.

This powerful, elegiac discourse on the part of our world Rachel Carson loved the most is a testament to the glory of the world we stand to lose. Carson's fascination with and reverence for her subject shine through page after page and inspire the reader to look closer and see more in the same ways she does.

Herman, Arthur. *Freedom's Forge: How American Business Produced Victory in World War II*. New York: Random House, 2013.

This book is perhaps an odd choice to make among my most recommended books about climate change, because I do not believe the words "climate change" even appear in the text at all. However, when we consider the response that must be undertaken to meet the challenge climate change presents, we struggle for historical precedents, particularly for ones in which we prevailed, and *Freedom's Forge* provides this example in gripping prose. The inspiring story of how the United States transformed itself from epically unprepared for the challenge it faced to simultaneously fighting and winning across two fronts, and more importantly of the path it chose to get there, makes this one of the most relevant books of all on the topic.

Larson, Erik, and Isaac Cline. *Isaac's Storm: A Man, A Time, and the Deadliest Hurricane in History*. New York: Crown Publishers, 1999.

Each summer, the United States is visited by a new crop of hurricanes, and each storm that comes ashore leaves swaths of the country cowering in the dark as the wind and rain rage overhead. No book so transports you into the heart of nature's fury as completely as this, while also recounting an epic tragedy of the man in charge of weather forecasting who got it wrong. We would do well to reread *Isaac's Storm* from time to time to remind ourselves of how completely we remain at Mother Nature's mercy, whether enduring a hurricane, forest fire, extreme rain event, or dangerous heat wave.

Oreskes, Naomi, and Erik M. Conway. *Collapse of Western Civilization: A View from the Future*. New York: Columbia University Press, 2014.

> This is the first of the books I recommend to those interested in learning how to make sense of climate change. Incredibly short but also highly impactful and loaded with facts, ideas, and theories, this book provides a view on the subject like no other. While quite fantastic in scope—it is an imagined academic treatise written 300 years in the future about how humanity failed in the twentieth and twenty-first centuries to correct course on climate change—it is entirely grounded in climate science, written by two leading authorities on the subject.

Perlin, John. *From Space to Earth: The Story of Solar Electricity*. Cambridge, MA: Harvard University Press, 2002.

> If indeed it comes to pass that the future of our species will ultimately be written through the lens of climate change, the story of clean energy technologies like solar and wind will be important to reclaim. It is surprising how little is available on these subjects to date. Perlin's lively and engaging telling of the way this technology came to be is not only exhaustive but also one of the only sources to synthesize and trace the whole amazing chronology of events and was essential to me in writing my book.

Wright-Gidley, Jodi, and Jennifer Marines. *Galveston, A City on Stilts*. Mount Pleasant, SC: Arcadia Publishing, 2008.

> As perhaps the nation's first scale geoengineering project to

attempt to manage Mother Nature's wrath, Galveston's reconstruction as a much higher city defended by a 17-feet-high seawall is a fascinating subject. This book centers on a rediscovered trove of photographs taken while the raising of the grade occurred, and the pictures tell the story of the titanic process much better than words can. If seawalls and grade raising are going to be features of our future, it may be that looking back on the ingenuity and methods of these turn of the (twentieth) century innovators would be time well spent.

INDEX